P & A Wood

Rolls-Royce and Bentley Heritage Dealers

Sales, Service, Repairs, Spare Parts and Complete Restorations

"Attention to Detail"

P & A Wood, Great Easton, Dunmow, Essex CM6 2HD, England
Telephone: 01371 870848   Fax: 01371 870810
E-mail: enquiries@pa-wood.co.uk   www.pa-wood.co.uk

Contents

The 30th Anniversary Edition

6 Contributors

7 Introduction

10 The Story of *The Automobile*
Co-founder Paul Skilleter and Publisher Douglas Blain chart our history

16 The Secret Kingdom of Oily Rag
Paul d'Orléans examines our fascination with leaving well alone

22 The Old Car Market, 1982-2012
Buying and selling is like a drug, says David Howard, and he's still hooked

26 Motoring Research, 1982-2012
The past is further away, but exploring it is easier, says David Burgess-W

32 Thirty Years of Finds & Discoveries
Michael Worthington-Williams reveals his all-time marmalade-droppers

42 Carry on Cartooning
Bill Stott brilliantly sends up the old-car world. Jon Dudley spoke to him

Features

48 Getting Started
Car won't start? At least it gets the brain going, says Gerry Michelmore

56 The Prop Half
Power boating and motoring have much in common, finds Hiram P Mate

64 FE and FENC: the Isotta Fraschini voiturettes
Simon Moore unravels the story of these fabled cars, and of four survivors

76 The First Motor Museum, a centenary
Michael Ware explores a collection first exhibited in Oxford Street in 1912

30th Anniversary Edition © Enthusiast Publishing Ltd ISBN 978-0-9674588-0-2

Editor
Jonathan Rishton

Editorial Assistant
Tom Grundy

Art Direction
Jon Dudley

Advertisement Manager
Peter Bromley

Accounts Manager
Caroline Paymayesh

General Manager
Jules Clifton

Publisher
Douglas Blain

Founding Director
Brenda Hart

Distributed by
Marketforce UK Ltd

Circulation Management by
Intermedia Brand Marketing Ltd

Pre-Press by
Zerofiftyone

Printed by
Wyndeham

Published by
Enthusiast Publishing Ltd,
PO Box 153, Cranleigh, Surrey GU6 8ZL
Tel: 01483 268818; Fax: 01483 268993
Email: enquiries@theautomobile.co.uk
Website: www.theautomobile.co.uk

The views expressed by contributors or readers are not necessarily those of the publishers. Every care is taken to ensure that the contents of the magazine are accurate, but the publishers cannot assume responsibility for errors. Whilst reasonable care is taken in accepting bona fide advertisements, the publishers cannot accept responsibility for accuracy or for any resulting unsatisfactory transactions. Unsolicited articles, photographs and transparencies are submitted entirely at the owner's risk, and the publishers accept no responsibility for loss or damage. The Editor retains the right to shorten, correct, re-cast or otherwise edit contributions including letters in the interests of clarity or accuracy, or for any other reason.

All material (including advertisements) generated by us in The Automobile is copyright protected and reproduction is strictly forbidden

Content

84 Marseille Maverick
The madcap adventures of Henri Rougier, by Reg Winstone

96 1914-18 RFC Crossleys
John Warburton salutes the four-wheeled heroes of WW1

106 Supreme Steamer
Thought steam lost its puff? Think again, says Sandy Skinner

114 Carburetters to Compressors
Karl Ludvigsen pays tribute to René Cozette's winning designs

124 The Phantom of Love
David Burgess-Wise on the Rolls-Royce fit for Marie Antoinette

134 Capolavoro da Corsa
Ronald 'Steady' Barker recounts his love affair with Lancia

142 Symphony: a very special MG TA
Park Ward only bodied big cars. Wrong, says Rutger Booy

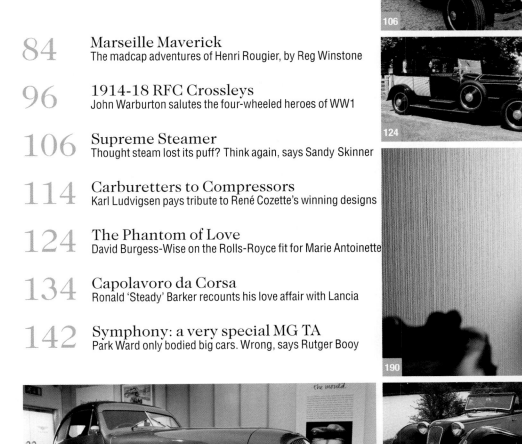

152 The Lincoln-Zephyr
Jonathan Rishton looks at the genesis of this mid-market V-12 saloon

162 The Embiricos Bentley
Brian Sewell assesses an extraordinary early experiment in streamlining

172 Goodwood 1948: the first meeting
A 16-year-old David Venables was there. 64 years on, he recalls the big day

180 Elliptical Excellence
Saab broke the mould with the teardrop-shaped Ursaab, says Lance Cole

190 Auto-biography: Sir Stirling Moss
Matthew Bell pays a visit to arguably Britain's greatest ever racing driver

196 A Grand Reunion
Count Marzotto was reunited with his 166 Ferrari. Keith Bluemel was there

204 The Beast of Turin
A 28-litre FIAT is nearly back on the road. Stefan Marjoram can't wait

Contributors

'Apsley'
An illustrator whose work has featured in the VSCC *Bulletin*, earlier this year Apsley collaborated on an article about Gerry Michelmore's Carden cyclecar. The pair join forces again in this issue

Hiram P Matelot
A first-time contributor to *The Automobile*, who shares his unparalleled knowledge of the early boating and motoring worlds in an article charting the crossover between the two disciplines

Ronald 'Steady' Barker
Renowned motoring journalist, with the largest stock of jokes and anecdotes of anyone we know, Steady Barker writes about the Lancia Lambda in this 30th anniversary issue

Gerry Michelmore
A hands-on engineer who lives on a steam launch on the Thames, Gerry restores and maintains Vintage cars when he's not fixing his own 1921 Carden cyclecar

Matthew Bell
A young journalist who grew up surrounded by Vintage cars, Matthew Bell writes Full Chat, a monthly column in *The Automobile*, as well as a regular series of interviews with old-car personalities

Simon Moore
The leading expert on Alfa Romeo's prewar eight-cylinder models; his books, *The Immortal 2.9* and *The Legendary 2.3*, are standard works on the subject. In this issue he turns his attention to another Italian marque

Douglas Blain
Douglas Blain edited his first motoring magazine at the age of 19. Editor and later co-owner of *Car*, he became Publisher of *The Automobile* in 1997

Paul d'Orléans
Globetrotting Publisher of The Vintagent, an acclaimed blog about old motorcycles, as well as Editor of Oily Rag, an online collaboration with *The Automobile* advocating the 'leave well enough alone' philosophy

Keith Bluemel
An renowned historian of the Ferrari marque whose expertise is sought worldwide, Keith Bluemel was able to reunite Count Giannino Marzotto with his Mille Miglia-winning Ferrari 166

Jonathan Rishton
Current Editor of *The Automobile* and of this 30th Anniversary Edition, in this issue he writes about the well preserved 1937 Lincoln-Zephyr in the magazine's own collection

Rutger Booy
Editor of PreWarCar, our sister website, Rutger Booy is a dedicated MG enthusiast who jumped at the chance to write about the Park Ward-bodied TA owned by his fellow Dutchman, Hemmo de Groot

Brian Sewell
Art critic of the London *Evening Standard* and renowned for his trenchant views, Brian Sewell is also a committed car enthusiast. In this issue he turns his trained eye on the famous Embiricos Bentley

David Burgess-Wise
David Burgess-Wise had a journalistic background before joining Ford's Corporate History Office. Since he retired from Ford he has had a prolific output as journalist, author, editor and translator

Paul Skilleter
One of the founders of *The Automobile* back in 1982, and now a publisher of books on Jaguar and other subjects, Paul Skilleter looks back at the magazine's early years

Nick Clements
Fashion photographer and founder of cult magazine *Men's File*, Nick Clements joined us for our Oily Rag Run earlier this year and captured our 1937 Lincoln-Zephyr on camera

Sandy Skinner
Erudite correspondent to *The Automobile*, Sandy Skinner makes the most of his amazing library. The meticulous rebuild of a Staride 500 is currently occupying much of his time

Lance Cole
An acknowledged expert on Saab, about which he writes in this issue, Lance Cole has also recently written a well received biography of Beverley Shenstone, the chief aerodynamicist on the Spitfire

Bill Stott
The Automobile's cartoonist in residence, whose monthly offerings capture perfectly the whys and wherefores of the old-car world. A selection of his best works is shown in this issue

Jon Dudley
As well as running the design agency responsible for the production of *The Automobile* each month, Jon Dudley is a committed user of Vintage cars and motorcycles

David Venables
Following his retirement from the post of Official Solicitor, David Venables has been researching many aspects of motor racing history. His most recent book is *Bentley: A Racing History*

Tom Grundy
Editorial Assistant of *The Automobile*, Tom is also a freelance film maker and photographer. His work features regularly in the magazine and on our ever-expanding website

John Warburton
Editor of the VSCC *Bulletin* from 2001 to 2008, John is an enthusiastic researcher and writer who has been restoring and driving Vintage cars for more than 50 years

David Howard
The Automobile's continental correspondent for many years, David Howard sets out from his home in the Loire Valley to report on auctions all over Europe with characteristic humour and insight

Michael Ware
Formerly curator of the National Motor Museum, Beaulieu, Michael Ware is a well known motoring historian and author who writes frequently for *The Automobile*

Karl Ludvigsen
A prolific writer and motor industry expert, Karl Ludvigsen's credentials are impeccable. In this issue we are treated to an extract from his forthcoming book on the history of supercharging

Reg Winstone
A freelance writer by profession, on topics less engaging than appear in these pages, Reg Winstone has a penchant for French machinery (and Voisin in particular)

Stefan Marjoram
A creative director at Aardman Animations, Stefan Marjoram is currently on a sabbatical to document the progress of the *Bloodhound SSC* LSR project while also indulging his passion for old cars

Michael Worthington-Williams
M W-W has written for *The Automobile* since the very issue. His monthly Finds & Discoveries and Automobilia columns are eagerly read the world over, as are his in-depth marque features

Introduction

My birthday came a day early in 2008. It should have been on 2nd September, but the day before I was asked: would I like to edit *The Automobile*? It was like being handed the keys to the Embiricos Bentley – an opportunity anybody in their right mind would jump at. But, just like getting behind the wheel of a one-off and highly-prized car, it's something you really don't want to get wrong.

The Automobile is a unique publication in a crowded market. It is the only monthly magazine to dedicate itself to pre-1960 vehicles, putting an emphasis on the quality of its writing and presentation. If an hour is a long time in politics, 30 years in magazine publishing is actually not that long. We are a mere whipper-snapper compared to many of our rivals, and yet, I feel, we more than match them in style and content. Certainly, I was well aware of the authority and high regard in which *The Automobile* is held.

Long before I took over, the magazine had found the formula for which it is so admired: that is, to provide thoughtful and thoroughly-researched articles about the kinds of machines you or I would want to read and, more to the point, drive and own. Attention to detail has always been key. I remember a close friend, on learning that I had accepted the job, warning me that *The Automobile* has the staunchest readership of any magazine, motoring or otherwise, that he knew.

It was a terrifying warning, and even now I open every piece of correspondence with trepidation. I have been reprimanded in writing and in person when readers have felt I've got it wrong. On the other hand, in a world where it's much easier to criticise than praise, we have also received a gratifying number of compliments.

And that is what makes *The Automobile* a success – that its readers care as much about it as the people who write it. We may be a tiny editorial team, but we have access to a vast amount of knowledge through our regular contributors and our readership. In the 30 years since the first issue appeared, information has become much easier to come by. Back then, the printed word ruled supreme: it did not have to compete with millions of pages of information available at the click of a button.

And yet, unlike so many other magazines and newspapers, *The Automobile* has not suffered from the arrival of the internet. Rather, we have benefited, as a thoroughly researched publication our readers can trust. Of course, we have also embraced the technological revolution, expanding and improving our website and forging a partnership with PreWarCar, a website that, we feel, shares our values. And the magazine itself has changed. In March, 2011, we enlarged the page size and the number of pages, as well as switching to a square-backed binding, silk paper and a revised layout. Change will always divide opinion, but it can also inject new life and excitement. The figures would certainly suggest it was welcome: unlike most other print publications, our news stand circulation is rising. The number of subscriptions is at its highest, as is advertising revenue. We are approached regularly by eminent historians whose groundbreaking research has no other monthly outlet – to such an extent, in fact, that we already have on file another two years' worth of first-class material.

A large part of *The Automobile*'s success is down to the energy and vision of our Publisher, Douglas Blain. He edited his first motoring magazine when he was just 19, which makes the 24-year-old he appointed back in 2008 look like a veteran. But taking risks is in Douglas's nature, and has served him well. His metamorphosis of *Small Car & Mini Owner* into the iconoclastic *Car* of the 1960s, made distinctive by it graphical boldness and brilliant writing from a relatively small band of trusted contributors, has been an influence on *The Automobile*: since he took over in 1997, his control has been transformative here, too. If it's not immodest to say so, he has a knack of hiring the right people and letting them get on with the job, free from excessive proprietorial influence but being there when needed, and the magazine has a punchier, less stuffy outlook.

I was not even born when *The Automobile* first hit the news stands, but I, and many others of my generation, share the same passion for using old machinery as it was intended, and learning why things were done the way they were.

Just over four years on, it's not for me to judge my performance at the wheel. But I do know that driving *The Automobile* continues to be an endlessly pleasurable and demanding experience. And, if you will forgive me one last outing of the motoring metaphor, it's not just the driver who wins the race. I couldn't do this job without the tireless efforts of Jon Dudley and his team, Jules Clifton, Brenda Hart, Caroline Paymayesh, Peter Bromley and others who make it happen. But above all I would like to thank you, the readers, for your continued support. I hope you will join me in celebrating *The Automobile* as it reaches this milestone. Here's to the next 30 years. **J R**

World Leaders in the Sale of Collectors' Motor Cars, Motorcycles & Automobilia

The motoring department at Bonhams congratulates The Automobile for providing 30 years of fascinating reading and imagery.
Long may this excellent publication continue to educate and inspire.

The Single-Seat Birkin Team Car No.1,
Brooklands lap record holder at 137.96 mph
1929 Bentley 4½-Litre Supercharged Racing Single-Seater
Sold £5.04m, Goodwood Festival of Speed 2012

Bonhams Motoring Department Sales Calendar Winter 2012/2013

3 December 2012	The December Sale Important Collectors' Motor Cars and Automobilia Mercedes-Benz World, London, UK
17 January 2013	The Scottsdale Auction Arizona, USA
7 February 2013	Les Grandes Marques du Monde au Grand Palais Exceptional Motor Cars, Motorcycles and Automobilia Paris, France
23 February 2013	Boca Raton Concours d'Elegance: Collectors' Motor Cars & Automobilia Florida, USA
13 April 2013	Classic California: Collectors' Motorcycles, Motorcars, & Related Memorabilia Los Angeles, California, USA

Bonhams can boast an unrivalled portfolio of blue riband auctions, from Scottsdale and Quail Lodge in the USA, to Monaco and Paris in mainland Europe, and Goodwood, Beaulieu, Aston Martin and Mercedes-Benz World in the UK.

To take advantage of what Bonhams can offer, please contact:

Motor Cars
+44 (0) 20 7468 5801
ukcars@bonhams.com

Motorcycles
+44 (0) 8700 273 616
motorcycles@bonhams.com

Automobilia
+44 (0) 8700 273 617
automobilia@bonhams.com

Bonhams 1793

"The Corgi" 1912 Rolls-Royce 40/50hp Limousine by Barker
£5.01m, Goodwood Festival of Speed 2012

1925 Bentley 3-Litre Speed Model Sports Two-Seater
£223,360 Goodwood Revival Sale 2012

1955 Lancia Aurelia B24 Spider Coachwork by Pinin Farina
£373,600, Goodwood Revival Sale 2012

The Ex-Works, Le Mans 24-Hours 1953-55 Austin-Healey Special Test Car/100S
£858,600, The December Sale 2011

1991 Aston Martin DB4GT Zagato Sanction II Coupé
£1.26m, The Aston Martin Sale 2012

The ex-Works/Lord Selsdon & Lord Waleran 1939 Lagonda V12 Le Mans Team Car
£1.31m, Goodwood Revival Sale 2012

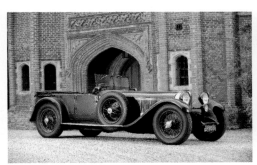
1928 Mercedes-Benz 36/220 6.8-litre S-Type Four-Seat Open Tourer by Cadogan Motors Ltd
£2.86m, Goodwood Revival Sale 2012

1965 Ferrari 275GTS Spyder Coachwork by Pininfarina
£573,800, Goodwood Festival of Speed 2012

1904 Delaugère et Clayette 24hp Four-Cylinder Side-Entrance Tonneau
£225,500, Veteran Car Sale 2012

Ex-Basil Dean 1932 Aston Martin 1½ Liter Le Mans 2-4 Seater Coachwork by E. Bertelli Ltd.
$208,500, Philadelphia 2012

1935 Riley 12/4 Kestrel Special
£61,900 Beaulieu 2012

'Floretta' – the Ex-works, ex-Wil-de-Gose 1908 Itala Grand Prix Car
£2.86m, Goodwood Revival Sale 2012

International Auctioneers and Valuers - bonhams.com/cars
Values stated include buyer's premium. Details can be found at bonhams.com

The story of *The Automobile*

The story of The Automobile

The first 30 years

Paul Skilleter, one of the founders, and Douglas Blain, the current Publisher, look back at the magazine's history

1. Michael Brisby, Editor of *The Automobile* 1982-1987

2. Peter Wallage (1987-1989)

3. Malcolm Jeal (1989-1992)

4. Brian Heath (1992-2003)

5. Michael Bowler (2003-2008)

The *Automobile* was one of those things which seemed a good idea at the time – and in due course proved to be exactly that.

Going back 30 years, the concept of a magazine which would cater primarily for prewar cars was that of Michael Shannon Brisby, a rangy, bearded, pipe-smoking Scot with a quizzical eye and an encyclopaedic knowledge of Vintage cars. I first encountered Michael when we both briefly worked for a car restoration company in Birmingham in the mid-1970s. Michael had been a local newspaper journalist in Scotland but, unusually, had then trained as a professional welder. When, in the late 1970s, I conceived the DIY magazine *Practical Classics*, I remembered this unique blend of talents and we hired Michael as the new magazine's first editor. Being both writer and restorer he brought credibility to the title and it was a big success.

Nevertheless, Michael was always slightly uncomfortable in the *Practical Classics* world of MGBs and Morris Minors. He owned an OM and was really a Vintage man at heart, so his suggestion that we should augment *Practical Classics* with a new magazine devoted to Vintage and PVT cars was not a surprise.

Lord Montagu's affectionately-remembered *Veteran and Vintage* magazine had long gone and there did seem to be a gap in the market. So, as we could promote the venture through PC at little cost, we at PPG Publishing took the plunge and in December, 1982, the first issue of *The Automobile* appeared. Its content and approach conformed broadly to the magazine you know today, with a spread of cars from Veteran through Vintage to the PVT era, sometimes straying into the 1940s but with a bias, shall we say, to the more esoteric end of the market. That reflected Michael's tastes. Indeed, the very first issue featured articles on that year's Brighton Run, M W-W's Angus-Sanderson, an ADM III Austro-Daimler, the Jaguar SS100 and an Edwardian Renault.

I believe our intention was to publish *The Automobile* quarterly and then, when we had established demand, progress in steps to publishing six then 12 times a year. This meant that the frequency of the early issues was irregular, before in May, 1983, it settled down to monthly appearances. The Finds & Discoveries column, which has remained one of the most popular features of the magazine ever since, began in the very first issue. Since May, 1987, it has been under the sole authorship of Michael Worthington-Williams.

It is fair to say that *The Automobile* was very

The articles in the very first issue set the tone for the scope and quality of features ever since

favourably received by the readership at which it was aimed. The only other bookstall magazine which covered the *Veteran and Vintage* scene to any extent was *Motor Sport*, and then necessarily in only a few pages. In fact Bill Boddy became a contributor to and a big fan of *The Automobile*, and I was similarly proud when another legendary motoring writer, Michael Sedgwick, also put pen to paper for us, although not for long as sadly he died in 1983. I did photoshoots for the magazine and occasionally contributed articles myself, finding it an enjoyable change from the 1950s and '60s machinery I usually wrote about – though I can claim some prewar credentials as my first car was a 1935 Riley Falcon and in the 1970s we had run a Morris Eight of similar age.

However, by 1986 our tiny publishing company was beginning to struggle and the figures appeared to show that *The Automobile* was eating into the profits made by *Practical Classics*. To be frank, there was a difference of opinion amongst us three partners about what should be done with *The Automobile*. Peter Hart, who had always solidly maintained faith in the magazine, chose to leave PPG Publishing and take it with him as part of the settlement. So 1987 began a new era for *The Automobile* under its new management – and, of course, Peter's judgement has been proved correct. The magazine continued successfully into the new century and flourishes today.

One result of this change was that Michael Brisby left and I undertook caretaker editorship of the magazine for four issues (January to April, 1987) pending its transfer to its new owners, Enthusiast Publishing under the leadership of Peter Hart. Michael then worked for Stanley Mann for several years – not least, I reckon, because he was adept at silently changing gear on Vintage Bentleys while demonstrating them to customers, some of whom, I suspect, later found that cog-swapping on such machines was not always as easy as Michael made it appear. In fact he tells me it was actually driving the cars that he found to be one of the most enjoyable aspects of editing *The Automobile*, along with attending weekend Vintage events as we all still do today.

After his time at Stanley Mann's, Michael then joined the Home Office where he worked as an investigating officer. Unfortunately that came to an end when in 2002 he suffered a stroke, and after some years rehabilitation – much aided by his wife Yasmin, who is a nurse – the family moved in 2007 to Dumfries, Scotland. I learnt this when after an interval of perhaps two decades I managed, with the help of M W-W, to track down Michael and Yasmin there. Michael has difficulty speaking but still has a great passion for old cars. The OM went a few years ago, but he now has a 1921 Model T Ford for which he built a body, so he definitely retains his restoration skills. He takes the Ford to local events. When I telephoned in October he and Yasmin had just returned from their annual visit to Hershey, staying with a good friend in New Jersey who restores Mercer and Stutz cars – right up Michael's street. Michael also has a talent I didn't know about: painting in oils. A member of the Fine Arts Society, he has sold a couple of paintings depicting old Napiers, although he is equally keen on sea- and landscapes.

I am sure everyone, especially long-term readers of *The Automobile*, will be pleased to know that the magazine's founding editor maintains his interest in the right sort of cars and, if not quite unscathed, is living happily in retirement with his family, Vintage car and dog.

Paul Skilleter

With the split from PPG, Peter Hart set up Enthusiast Publishing in May, 1987, and hired Peter Wallage as the next editor of *The Automobile*. He remained in the post until April, 1989, when he was succeeded by Malcolm Jeal, a respected historian who set new, scholarly standards for the magazine. In May, 1992, he was replaced by Brian Heath, the longest-serving editor, who stayed until October, 2003.

The story of The Automobile

Covers have become simpler, less wordy and more uniform over the years but, underneath, The Automobile still tills the same field and appeals to the same market as when it started 30 years ago

I had been talking for some time to Lionel Burrell, an old friend from my days as editor and later co-publisher of *Car* magazine, about getting involved again. Some 20 years had gone by since I parted company with *Car*. That was at the height of the original Middle Eastern oil crisis and I remember feeling bad about blasting all over Europe, averaging 75mph and 10mpg, in exotic Ferraris and Lamborghinis while ordinary mortals agonised over where the next tankful was coming from.

My solution at the time had been to buy an elegant horse-drawn landaulet and install it, by gracious permission of Her Majesty, in the Royal Mews. By hiring my horse and coachman out to Fortnum and Mason by day for prestigious deliveries, I was able to use my own equipage for evening expeditions to the opera.

But by the mid-1990s I was beginning to think it was time to return to my motoring past, deeply rooted as it was in the Vintage tradition (my first proper car had been a Bentley Three Litre, bought in Sydney for £200). That path led to Peter Hart's door in leafy Cobham, Surrey. Word had reached me via Lionel Burrell that Peter was thinking of retirement. We hammered out a deal and I took his place in the Publisher's chair in time to oversee the February, 1997, issue with Lionel as my art director and Peter's wife, Brenda, keeping a beady eye on the office.

Brian Heath was editor then, but by late 2003 he, too, had begun to think of retirement. Lionel's friendship with Michael Bowler proved a godsend at this juncture. They had worked together at *Classic Cars* for many years, after which Michael, a qualified engineer, had gone on to become technical director of Aston Martin. He jumped at the chance of getting back in the journalistic saddle, reminding me that Vintage cars were in his blood, his late father having been one of the early stalwarts of the VSCC.

Michael and Lionel both stepped down in 2008, leaving the door open for an altogether younger team with fresh ideas. Jonathan Rishton, then barely 25, had been working with my old friend Steady Barker, who recommended him to me as "the next William Boddy". Working with Lionel's former assistant, Rob Pigott, Jonathan with my enthusiastic support began taking the magazine steeply upmarket, a process which culminated in what is termed in the trade a 're-launch' in March, 2011.

That was when *The Automobile* acquired its distinctive unvarnished stiff covers, silk-textured paper and square-backed binding. Jonathan, together with Jon Dudley, who took over from Rob after the latter moved to France, has since dragged us well and truly into the 21st century with a new website, an Oily Rag blog, a new distributor, a digital edition and even our own event: the increasingly popular Oily Rag Run.

Comparing the latest issue with our very first edition of 30 years ago, it is clear we have come a long way, covering this vast subject now in much greater depth and with very much more consistency, albeit without the early bias towards commercial vehicles. But there is a thread of continuity which ensures that many of our early readers are still with us. Unlike other motoring magazines, this is one which repays conservation and is referred to again and again. The result is that we have, amongst you all, a far greater proportion of regular subscribers than rival titles and a very substantial sale of binders.

I have always believed the way to judge a magazine's true worth is by studying its correspondence columns. Ours function like a forum for dedicated and incredibly knowledgeable enthusiasts from all over the world to swap ideas, pool opinions and substantiate or refute theories on every aspect of their hobby. As long as that appetite survives, and as long as there are connoisseurs who appreciate a publication of genuine quality, *The Automobile*, I believe, will continue to thrive.

DEB

BLOCKLEY TYRE COMPANY

Blockley Tyres congratulate The Automobile on their 30th Anniversary

The Right Appearance
The Right Performace
The Right Price

For Road.......and for Race!

Post War Cross Ply   Post War Radial   Pre War Cross Ply

Blockley Tubes
We now supply our own
Blockley Superior Tubes
available for all our tyre sizes
in both road and race type

!BEWARE OF IMITATIONS!

www.blockleytyre.com   Tel: 01386 701717

The secret kingdom of Oily Rag

We at Oily Rag celebrate elegant Parisian ladies of a certain age who defy the knife, the needle and the silicone plump in favour of just being themselves. This Facel Vega FV-2 convertible is exactly to whom we refer

The Oily Rag philosophy has a mixed relationship with barn-stored vehicles; we're sad they languish in obscurity, but are delighted to find, revive, and use them. Events such as *The Automobile*'s Oily Rag Run gather the like-minded for a pleasant day stretching the limbs of our barn finds

'Use them as the maker intended', or perhaps as the maker never imagined! A 40-year-old off-road Triumph TR5T remains eminently suitable for exploring the wilds of our world, and can take you places which veer deeply into the Epic

Nobody in their right mind would ford Utah streams on a restored Triumph TR6R Tiger, but a slightly weather-worn example doesn't mind a bit of moisture. Old motorbikes were terrific all-rounders, handling loose stuff with the same aplomb as tarmac; there were few specialised dirt bikes before the 1970s, and round-the-world trips were taken on street machines. They still work, and are still tremendous fun

Dennis Severs was a patron saint of Oily Rag, transforming his London home into a living museum of 19th century daily life, without electricity or modern convenience. Severs didn't drink the Kool-Aid; he invented the recipe! His purity of purpose went far beyond our usual embrace of 'utilising the old'; his entire world was built from the old. Having recently experienced a week in New York City without power, heat or water, I can confirm it takes a complete mind-shift to live in a candle-lit world. You can visit the Severs house in Spitalfields, London, and see for yourself

Where were you in '82?

1982, Rutger Booy in his tenth year of MGA ownership. Transporting Sinterklaas to the town of Oostvoorne.

1982, Joris Bergsma's first classic sports car: 1969 Alfa Romeo 'Graduate' Duetto Spider.

The editors of PreWarCar-PostWarClassic congratulate The Automobile with bringing automotive joy from 1982-2012.

www.PreWarCar.com   www.PostWarClassic.com

P.s. We are eager to learn what set of wheels you were taking care for exactly 30 years ago.
Please send photos showing your car & yourselves in 1982: info@prewarcar.com

The old-car market

David Howard looks back on his career in the motor trade and ponders on the advances of the last 30 years

I was born at the outset of World War Two with what was apparently a perfectly normal blood group, but one which I have since assumed to be liberally mixed with motor oil – possibly Castrol, Mobil or, most likely, Pratt's. From a terribly early age all I thought about was motors, even burying prewar Dinkies deep in the garden at home to protect them from the threat of the Luftwaffe.

As a teenager, answering the stupid educational demand for an essay on what I had done in the holidays, I would answer alternately 'breakers' yards' or 'scrap yards'. These places were heaven sent to me, and were where the ever-present urge to buy and sell set in. Austin Sevens for a fiver, Standards, Fords and Morris, all came cheap, towed home on bald tyres, fretted over, given what I hoped was a new lease of life and sold on. I didn't worry too much about the chaps (always chaps) who bought them; the welfare of the car was my priority.

Two memorable scrap yards stay imprinted on my mind. The Goodey Brothers' place at Twyford, to which I journeyed all day with my father's Fordson Major and a huge farm trailer to rescue a 1924 Wolseley 16hp two-seater (I think its registration started with XR). I painted it in red Valspar with black wings, and saw it many years later at a VSCC gathering. The second, two years later, was Smith's yard on the Foleshill Road, outside Coventry, where I spied a 1926 AC Six two-seater, registered EX 1945, looking very sad. I bought it for £25 and, with the aid of two friends and a Bean tourer, dragged it home. I was gratified this year to recognize it as Matthew Crawley's new car on the television programme, *Downton Abbey*.

A decade in the modern motor business did nothing to dull my appetite for 'proper' cars and I ran a Vintage offshoot to the business all that time. Little old ladies who came into my garages in the early '60s were driving cars that enthusiasts today term 'classics'. They would ask for their batteries, plural, to be checked – a throwback to the prewar pair of six volts.

It would be about this time that auctions for collector's cars began in earnest: Sotheby's and Christies at Beaulieu, Mike Carter at Alexandra Palace (before it burned down) and smaller events such as Husseys in Exeter market, to name but a few. We used to use a Champion spark plug at five bob and a Mars bar to test the temperature of the market; these two objects supposedly hardly increased over the years. Over the ensuing 40 to 50 years the prices of older cars have of course rocketed, although there have obviously been huge downturns in the market, as at the end of the '80s, when many 'investors' as opposed to enthusiasts caught well deserved colds.

Michael Ware, when Curator of the Montagu Motor Museum at Beaulieu, coined the clever word 'autojumble'; now autojumbles take place most weekends and autojumblers, active all the year round, are avid buyers at auctions. There was a time, not long ago, when number plates were the means of identification on the front and rear of vehicles, as they still are on the continent. Now, of course, in the UK they not only change hands at enormous prices, but have sparked off a whole new industry of 'personal plates'. These appear on usually more flashy motor cars to add to their owners' already over-inflated egos.

By the mid-'70s I was tired of the modern motor trade and concentrated solely on Veteran, Vintage and, to a degree, so-called classic cars. I have never been able to enlighten myself, or anyone else for that matter, as to what exactly constitutes a classic. Having shed some 20 staff and three garages, it was a welcome relief to work as a one-man band, with only my long-suffering wife to help in the office. I operated from converted farm buildings at my home in rural Hampshire. The union or brotherhood of Vintage car dealers was very close and many deals were done in the trade. We were, and in many cases still are, great friends.

Auction houses at the end of the '70s were few and far between. Christie's and Sotheby's still led the way, with the Hon Patrick Lindsay of Christie's perhaps the doyen of auctioneers and Malcolm Barber holding sway at Sotheby's. Smaller auction houses were beginning to spring up. Linden Alcock in the west, Walton and Hipkiss in the Midlands and British Car Auctions at Blackbushe opened a Vintage department, to name but a few. In London, Coy's branched out and Phillips started a car sales department. Later, in the '80s, Robert Brooks, who had for years been understudy to Patrick Lindsay at Christie's, came down to see me at Romsey for a pub lunch and asked my views on the possibility of him starting his own auction house. I seem to remember my advice, if it was worth anything, was "to give it a go". He soon absorbed Phillips and went on to take over Bonhams, forming the most influential old-car auction house of the present day.

I think one of the most romantic sales I ever attended was held by auctioneers Colliers, Bigwood and Bewlay, who sold off the vehicles of a company near Wolverhampton belonging to one Harry Ellard. He ran largely Morris Commercial lorries, but had Lagonda and Invicta cars for his own pleasure. When he had finished with each vehicle, instead of selling it on it was merely pushed deeper and deeper into the factory buildings. The cars were complete, covered with dust, and had very low recorded mileages. I remember attending the auction with my old friend Danny Margulies, doyen of the Queen's Gate Mews dealers, and I managed to buy a delightful 4½ Litre Invicta drophead which I subsequently sold, very

Research – but not as you knew it...

1. David Burgess-Wise lectures to a carefully chosen audience at the magnificently endowed Collier Collection in Florida

2. The collection at Beaulieu's motoring library includes engineering drawings, photographs and documents as well as books, brochures and magazines

3. The Aston Martin Heritage Trust is one of a select few of such organisations with its own curated premises and archive

4. This tiny disc contains every page of *Motor Sport* from 1924 to 1949

5. Inside the Collier Collection's Library, helpful and knowledgeable staff are on hand to help researchers

6. Even the most esoteric motoring ephemera is collected, catalogued and stored by leading libraries worldwide. This ultra-rare publication is accessible online, page by page

7. The number of motoring books kept at Beaulieu runs into the tens of thousands

8. The Society of Automotive Historians in Britain, to which many of *The Automobile*'s contributors belong, issues its own publications with fresh, scholarly research

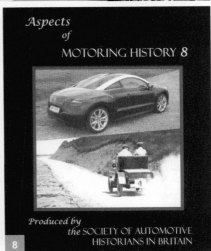

1. This extremely scarce publication, a souvenir of the 1907 Targa Florio, has been digitised by the Collier Collection and is now accessible online

2. Rare correspondence, such as this letter from C S Rolls, is preserved in archives all over the world and made available to researchers

shut the lid on the wreckage and tiptoed away with the ancient tome under my arm. Fortunately for my continued employment prospects, in those days photocopiers were a mystery that lived in their own isolated cupboards, doubtless because of the pervasive smell of the witches' brew of chemicals they contained. The young perpetrator of the sacrilege remained undetected.

The damage to bindings caused by repeated photocopying of bound volumes is another problem that modern technology has eradicated. Nowadays, excellent copies can be taken with a tiny digital camera without having to force the binding flat. I remember, too, that when I was researching the centenary history of Daimler that Lord Montagu and I wrote in 1995, certain old handwritten company ledgers were deemed too frail for photocopying. I had to transcribe great chunks of text in pencil, as pens of any sort were strictly forbidden anywhere near the ancient documents. If only digital cameras had been available then…

Of course, computers are essential to such operations. It's edifying to remember that 30 years ago we were still reliant on typewriters to transfer our thoughts to paper. When the first home computers did become available shortly afterwards, their word processing powers were minimal to say the least. I recall how pleased I was with my BBC-B computer, which I bought from a shoestring company operating out of a room behind a greengrocer's shop in Harlow New Town, yet it had no inbuilt memory and could only cope with 10 pages or so of text at a time. These I had to save on to 5.5in floppy discs and print out on a dot matrix printer that offered only one choice (not a very pleasing one) of typeface.

Nevertheless, I managed to churn out the text of many articles on it, though it offered the publisher no advantage over a typewriter as the printroom operatives still had to retype everything before it went to press. When clients wanted text on diskettes that their computers could read, I had to post my floppies to a company on the East Coast that had some magic software that would transform BBC-Basic into discs compatible with the client's system. For a subsequent job – a partwork encyclopaedia – the publisher provided a word processor on loan, an Amstrad with an annoying small green screen that gave me eyestrain. It was, like the BBC-B, incompatible with any other system, and I was very glad to give it back at the end of the commission. I gave Steady Barker, who obviously got on better with Mr Sugar's brainchild than I did, my leftover diskettes.

It was only when the first affordable PCs became available in the late 1980s that the real advantages of the home computer as a working tool became evident; the facility to store and recall raw research material was a real advantage, particularly when it came to writing books. Equally, the ability to move copy around without resorting to crude cut-and-paste methods really came into its own when making the final edit of a long book.

It was round then that the first optical character readers made their appearance. When I was working on the definitive history of Ford in the Henry Ford II era (the project was shelved *sine die* in the financial crisis at the beginning of the 1990s, when I decided to leave the company), Ford of Britain bought an early OCR scanner for around £15,000. Seeing the advantages of this for transcribing documents, I sent my secretary down to see the scanner people with a book from which I wanted a passage copied. She returned with a printed page of gobbledegook. "The scanner can only read typewritten documents," she lamented. As far as the Corporate History Project was concerned, that was £15,000 down the drain. Now, just about every home scanner comes complete with OCR facilities and can scan in several languages, a must when working on translations of foreign texts (though I have yet to find a reader that can cope with the black-letter German used in my Edwardian volumes of *Allgemeiner Automobil Zeitung*). So today the researcher has a wide array of tools that were unavailable when the first issue of *The Automobile* came off the presses – but what of the research facilities?

For one thing, original material is far more accessible than it was when I began writing about early cars. Not only have many major companies established their own archives, with Mercedes-Benz and Audi among the most active and the Fondation Berliet in Lyon working on a broader basis with voluminous holdings on the entire automotive industry of the Lyonnais area, but substantial amounts of material are held by educational establishments and county and municipal archives. Most of these are accessible online, and I have found them extremely helpful in dealing with the most obscure enquiries.

Several of the larger car clubs have their own archives dealing with the cars that are their special interest: the Aston Martin Owners' Club, the Bentley Drivers' Club and the Rolls-Royce Enthusiasts' Club are

Research – but not as you knew it...

The Automobile's popular website features a searchable database of contents from its 30 years of publication

among those that have set up charitable trusts to safeguard their holdings, while the library of the Veteran Car Club not only holds a wide range of books and magazines but also has the highly important Simms papers, recording in almost obsessive detail the early days of the motor industry in Britain.

Many of the major museums also have comprehensive archives. Our own National Motor Museum has a very helpful Motoring Research Service, with its library open by appointment from Tuesday to Friday, and the British Motor Industry Heritage Trust and Jaguar-Daimler Heritage Trust have large archival collections. Most of these may be visited, but remember that many may charge for research, whether carried out personally or by museum staff. Overseas there are major collections of archival research material held by establishments such as the Henry Ford Museum in Dearborn, Detroit, whose holdings include the archives of the late Henry Austin Clark. Other outstanding collections are those of the Biscaretti Museum in Turin and the Collier Collection in Florida.

The exemplary work carried out by the Collier Collection saw the establishment of an educational body called the Revs Institute in 2009 to guide and manage the Collier Collection Library and Archives. This is part of a 20-year development programme that includes a multidisciplinary academic link with Stanford University in California. Among the collection's substantial holdings, which number more than a million items and include the libraries of Karl Ludvigsen and the late Peter Richley, are more than 17,000 automotive books, 200,000 issues of automotive journals from around the world and more than 300,000 photographs, negatives, motion pictures and graphic illustrations, plus an historical ephemera collection of more than 400,000 items.

The Institute's links with Stanford move the study of automotive history onto an altogether more academic plane. 'The automobile is surprisingly under-studied by scholars,' says Professor Clifford Nass, the director of the Revs Program at Stanford, 'But this cultural icon is worthy of – and overdue for – deep understanding on every front. Our primary goal for the Revs Program at Stanford is to create a vital and much-deserved intellectual community around the car as technological and aesthetic artefact and cultural symbol.'

Such an attitude is a world away from how it was 30 years ago. I remember visiting the archives of one long-established company around that time in their near-deserted Victorian factory in the London suburbs and finding unique records going back more than a century randomly piled up in cupboards in a dusty storeroom. The 'archivist', who seemed to have no other function apart from being responsible for unlocking the door and making tea, had no idea of what was there and research was very much of the needle-in-a-haystack variety. Fortunately, I am naturally able to read at the rate of several thousand words a minute, which is a fantastic help in such circumstances.

The 21st century has brought new tools to the aid of the researcher that can bypass hours of patiently thumbing through bound volumes of old magazines. The publishers of *Motor Sport* have digitised the entire run of the magazine, with the golden years of 1924-49 – including the incredibly rare 1929 issues – available on a single, searchable CD. Here at *The Automobile* we are hard at work on a similar system. Some of the earliest American and French magazines have been digitised with various degrees of competence; I have *La Locomotion Automobile* for 1894-95 on a CD, but it's not searchable and no scrolling is available – it has to be laboriously viewed a spread at a time.

Usefully, for there is great crossover between the early years of motoring and aviation, every issue of *Flight* from No 1 in 1909 has been digitised in PDF form and placed on the internet. Seekers after the more obscure aspects of motor sport history can find the results of even quite minor Australian events by searching online archives of that country's regional newspapers, while many American newspapers have also digitised their early editions. And for aficionados of between-the-wars French concours, the archived issues of the society magazine *l'Officiel de la Mode* have been published on the internet.

Also on the internet are holdings of the US Patent Office and digitised copies of many early motoring books, all searchable by computer. A real treat – again a venture by the enterprising Collier Collection – is the 1907 issue of the Targa Florio souvenir book *Rapiditas*, a slightly-rarer-than-hen's-teeth gem from Peter Richley's library that also covers the associated voiturette and La Perla del Mediterraneo motorboat races, faithfully published on the Revs site, right down to the tissue interleaves protecting the art plates.

But while whole libraries of motoring literature are available on the internet, the basic rules of research still apply: diligence, curiosity and groundwork are needed for best results.

Finally and most usefully: since the birth of *The Automobile*, serious motoring historians in this country have joined together in their own organisation, the Society of Automotive Historians in Britain, the UK chapter of the American SAH (which was founded in 1969). It enables members to tap by telephone or email into the collective wisdom of their peers and, by publishing learned papers, elevates the business of automotive research to a more professional plane.

For myself, though, despite all the electronic aids now available, I have to say I still most enjoy the thrill of the chase in leafing through old books and magazines in search of that elusive fact that solves a century-old mystery or leads me down an entertaining blind alley. The printed word, I firmly believe, will never die.

JOIN 'THE RIGHT CROWD'

Congratulations to our successful Brooklands Racing Team in 2012

BROOKLANDS TRUST MEMBERS

Join Brooklands Trust Members, the official support organisation for Brooklands Museum and enjoy unlimited **FREE ENTRY*** to the historic Brooklands site.

Brooklands Trust Members is the only organisation dedicated to raising funds for the preservation of the site of the world's first purpose built motor circuit – The Birthplace of British Motorsport and Aviation – and to supporting the work of Brooklands Museum.

There are now over 10,000 Brooklands Trust Members. Join as an Individual (£30 pa), a Couple (£45 pa) or a Family (£55 pa) and enjoy *free entry** to Brooklands Museum, an extensive social programme and special members' events. Or upgrade to Club level membership (£85 pa) to enjoy even more special benefits, including use of the Members' Bar in the famous Brooklands Clubhouse.

Brooklands Museum has a wide range of motoring and aviation exhibits ranging from the 24-litre Napier Railton to a unique collection of Vickers/BAC built aircraft, including the only Concorde in South East England with public access.

Visit: www.brooklandsmembers.co.uk

For full details from the Brooklands Trust Members Administrator
call: 01932 857381
or email: info@brooklandsmuseum.com

BROOKLANDS MUSEUM

* During normal opening hours, may be subject to an extra premium for specific events

Thirty Years of Finds & Discoveries

Michael Worthington-Williams, who has written for every single issue of *The Automobile*, recalls some of his best discoveries

I once wrote that 'the traditional scrapyard is my spiritual home'. Over the past 50-odd years I have known many of them. In Sussex I often visited Harris's and Tom Brimfield's in Burgess Hill; Willett's in Noah's Ark Lane, Lindfield; Percy and son Richie Vokes in Adversane and Billingshurst (later Mutton's). Elsewhere I used to haunt Medlar's, Arkenstall's, Motolympia, Harry Buckland in the Golden Valley, Gordon Passey (licensed horse slaughterer – you needed a strong stomach there). Then there was Tan-y-Groes – also literally a knackers' yard, Watton Metals in Studley, a semi-derelict mansion near Bow in Devon surrounded by acres of hulks, and many, many others.

Before *The Automobile* was first published in December, 1982, I had already notched up hundreds of discoveries: a 1926 BSA motorcycle taxi in Brighton when I was 18, a 1923 Bentley Three Litre in Shifnal, Shropshire, two (yes, two) Hispano-Suiza K6s in a bricked-up railway arch in Brixton, an 8A Isotta-Fraschini town car by Castagna in an open-ended shed in Borth, near Aberystwyth, and a 1921 Rolls-Royce Silver Ghost in Kent, plus others too numerous to mention.

But my brief, in this 30th anniversary special edition, is to list the 10 most important discoveries which have featured in *The Automobile*. Given the number I have reported over three decades, this has been an almost impossible task. I have, therefore, decided that those chosen should be the ones which were the most important to me, which were the most historic, in terms of age, or which I should most have liked to own. The Editor has agreed that these may not necessarily have appeared in the Finds & Discoveries pages, but if they were genuine discoveries and have featured in the magazine they can be included.

In fact, the Finds & Discoveries column was not originated by me but by the magazine's first Editor, Michael Brisby. It was not until issue number seven in September, 1983, that I first passed on the discovery of remains of a Clément-Bayard to the column. Although I contributed regularly thereafter, it was not until May, 1987 – by which time I was writing all of it – that my byline first appeared, under the late Peter Wallage's editorship.

Angus-Sanderson

1. The Angus-Sanderson complete with crane when still in use, c1973

2. The Angus-Sanderson as acquired in 1977. The tin roof prevented tits from nesting in its garage

3. The same car following restoration, with M W-W outside Glaspant Manor

However, my first discovery for *The Automobile* appeared in that very first issue, and it was an extremely important one for me. Under the title For the Love of Angus, I recounted my love/hate relationship with my 1921 Angus-Sanderson 14hp tourer. My passion for this rare marque commenced in 1954 when, at the age of 16, I bought a set of Rankin Kennedy's *The Book of the Motor Car* for 15 shillings, which I paid in weekly instalments of five shillings from my 30-shilling weekly wage. On page 90 of Volume IV was a photograph of the 1918 prototype which, with its radiator designed by Cecil Kimber, I thought the most sleek and beautiful Vintage car I had ever seen.

My weekly forays around my then home in Sussex, searching barns and fields for wrecks and getting to know the local scrapyards, failed to turn up an example, but eventually I received a tip-off from David Burgess-Wise – now, of course, also a valued contributor to *The Automobile*. He directed me to the wood yard near Tenterden owned by George Jarvis, a nonagenarian ladder maker. There I was shown the 1921 Angus-Sanderson which his father had bought for £2 in a farm auction in 1937, following the death of its first owner, Edward Collison of Goldwell Farm, Biddenden. The car, which Mr Collison had kept in immaculate condition, was by 1937 obsolete and virtually worthless.

Mr Jarvis had removed all bodywork behind the scuttle, installed a Buick gearbox behind the original Wrigley unit to provide a double reduction, adapted the 'A' chassis frame from an Austin Seven to make an improvised crane, and removed the tyres, replacing them with wooden segments around which were shrunk iron tyres and to which iron cleats were welded on the rear wheels. The result was a crude tractor which, for the next 40 years, operated as a tool of trade in the wood yard and its surrounding 50 acres, pulling out tree stumps and occasionally being hired out to dredge ponds and other menial work.

It was not until some years later, however, that George was willing to sell. He was by then 97 and had been ordered by his doctor to give up driving because of failing eyesight. By that time the condition of the car was as shown in the photograph you see here. Attempts to discover the original registration mark proved fruitless until I received a letter from 88-year-old Alice Viggers, following an appeal in all Kent newspapers. Her late husband had been manager of Weekes & Son, of Perseverance Ironworks, Maidstone, who had sold Edward Collison all his farm machinery and the Angus-Sanderson.

Alice put me in touch with Edward's two daughters by his second marriage to a much younger wife, and they were able to provide me with a superb sepia photograph showing Edward and his youthful bride in the car on the day of delivery in 1921. The photograph was specially commissioned from High Street photographer Stickells of Cranbrook, and I still have it. It proved invaluable during the restoration which followed over a 23-year period – and which cost me a fortune.

Panhard

In October, 1983, and under the title A Trio of Primitives, I recounted the stories of the three most important discoveries I had made up until that time. I was then head of the Vintage vehicle department at Sotheby's (indeed, had been since 1976), and there is no doubt that that rôle opened many doors which would otherwise have been closed to me. The three cars described all pre-dated 1900, and two of them originated in Italy.

The first was a Panhard et Levassor dating from 1894 which had been discovered in a dry vault beneath a Florentine villa. It had been in a single family from new and had been off the road certainly since before the Great War, and probably since before 1900. Its owner was a proud aristocratic lady whose father had purchased the car new, direct from Panhard et Levassor. Panhard had, of course, had successful cars running as early as 1891, albeit with their Daimler engines mounted amidships. Not long afterwards, however, they adopted what we now know as the *système Panhard*, which was to become the industry norm for the following 50 years. This comprised a front-mounted engine, mid-mounted gearbox and driven rear wheels, and that was the layout of the Florentine car.

Doubtless our Italian nobleman had been aware of the Panhard which had been joint winner of the Paris-Rouen Trial of 1894, one of the first properly organised competitive events for motor cars. The other winner was a Peugeot which, like the Panhard, employed a Daimler engine. It is apparent, however, that it was not this event which influenced his purchase. Documents still in the family's possession disclosed that the order had actually been placed in 1893, and the car was generally believed to have been the first on the roads in Florence.

Letters from the first owner to the factory indicated that overheating had occurred, resulting in the fitting of an additional radiator by the works themselves. Otherwise the car was virtually unmodified when found, featuring its original hot-tube ignition, single chain drive, spoon brakes, iron tyres and unprotected gear pinions (closed gearboxes were not introduced until 1895). It is fitted with Daimler engine number 344. We do now know that a similar car with engine number 227 was delivered in November, 1893, so the car may be older than 1894, although the latter date seems most likely.

Initially problems were encountered in obtaining permission from the Italian authorities for the car's export licence, but when no buyer in Italy could be found who was prepared to pay the asking price, and because the car was not of Italian manufacture, this hurdle was overcome. I was able to persuade HM Customs that the car could come in under the antiques tariff 97.05 as a collector's piece of historic interest, paying no VAT or duty. It was eventually sold by Sotheby's to the Caister Castle Motor Museum, where I believe it is still on display.

1. The 1894 Panhard et Levassor found under a villa in Florence

2. Note the auxiliary radiator at the front

Bugatti

1. 1938 Bugatti Type 57 Atlante coupé

2. The Bugatti Stelvio by Gangloff. It harboured a very dead rat

No report on exotic discoveries would be complete without a Bugatti, and over the years I've reported quite a few. In November, 1987, Sotheby's sold a number of unrestored cars from the collection of the late Kenneth Ullyet, JP, including a Bugatti Type 57 Atalante coupé. This prompted Paul Jaye to tell me of another derelict Bugatti he had acquired – a 1935 Stelvio cabriolet by Gangloff. You see it here as acquired, following a somewhat chequered history.

It was not included in Barry Eaglesfield's Bugatti book, published in the '50s, despite having passed through the hands in 1954 of Brussels dealer de Dobbeleer, who had assisted Barry with the book. De Dobbeleer had acquired it from the original owner and sold it to a Lt Colonel Beresford of the US Army, who took it back to the USA with him. He ran it for a few years before a slipping clutch and water leaks caused it to be laid up.

It remained in dry store in Texas for the following 25 years until the Colonel's death. It was then bought by British dealer Brian Classic, then Dan Margulies and finally Paul. The engine was found to be turning freely but the hood, upholstery and trim had all suffered mightily from the ministrations of a family of very large rats, one of which had evaded Customs and quarantine regulations by coming in with the car and expiring within. When I made my inspection the smell of corruption was still powerful, but restoration followed.

AUTOMOBIL VETERANEN CLUB AUSTRIA
AUSTRIA HISTORIC
29.06. - 07.07.2013
Southern Route
Crossing Austria from West to East

Travelling along a great number of the most beautiful country roads of Austria, climbing some breathtaking pass routes, through picturesque valleys, along quiet and famous lakes, visiting sights of historic and cultural excellence.

No scores !
No official speaches !
No sponsor obligations !
Sheer driving pleasure !
The route is the challenge !

Total distance approx. 1.200 kms,
max. gradients 15%
Daily routes approx. 200 kms
Long, demanding, challenging, feasible

4**** hotel accommodation
Daily luggage transport to hotel
service support vehicle
for all pre 1945 cars
limited to 50 cars in total

For more information contact:
AVCA, POB 332, A-1012 Vienna
mailto: office@avca.at   or   www.avca.at

"First come - first served"
Our Northern Route 2012 was fully booked
as early as seven months before the start.

P.s.: This rally won't be an easy one
but it will be of memorable beauty -
a demanding but feasible challenge for man and machine.
Hence, good overall condition of both driver and vehicle will be of advantage !

Paris-Madrid/
24-26 May/2013/

A Classic Reliability Trial to celebrate the 1903 Paris-Madrid Race for the Charles Jarrott Trophy

To find out more about our rallies call **+44 (0)1252 717175**
email **info@hhclassicrallies.com** or visit **www.hhclassicrallies.com**

Carry on Cartooning

Jon Dudley speaks to Bill Stott, *The Automobile*'s cartoonist-in-residence

One of the many delights upon receiving a new issue of *The Automobile* is the monthly cartoon. Always bang on target, Bill Stott's humorous observations of the old-car world and its inhabitants continue to delight. How does he do it: more than 20 years of coming up with clever ideas, month after month? In this, our anniversary issue, we thought it was time to talk to the man behind the drawings.

Bill and his partner live deep in rural Cheshire; he himself was born in Preston. He attended art college there and subsequently became a teacher, supplementing his income by drawing cartoons for which he found he'd got a real talent. Despite failing his 11-Plus he rose to the heights of HM School Inspector but, becoming disillusioned with the attendant bureaucracy, returned to the classroom. Having retired from teaching aged 50, he's now 68 and "As most old people say, I don't feel it – then I have to spend 30 minutes looking for my glasses." He enjoys his bucolic life immensely and seems to be a very happy man. Alongside his cartooning he produces 'serious' non-figurative painting, although quite rightly he regards cartooning as a legitimate art form, too.

Working fast, Bill can turn round a cartoon for *The Automobile* in something like two hours – an incredible rate to someone like me who struggles to draw convincingly. Years of experience and a brisk creative mind are, he says, the keys to his success and productivity. He acknowledges a debt to local hero and fellow cartoonist Bill Tidy, and they appear to have sprung from the same mould: attention to detail is a shared characteristic, and the bolt-eyed moustachioed buffers are favourite characters portrayed by both men. There's something of H M Bateman in their DNA.

Bill draws for a diverse group of publications as well as *The Automobile*, and on a variety of topics, too, from stationary engines to philosophy. There's humour in most subjects, he says, but it seems that in the old-car world he's struck a particularly rich vein. We can all identify with his characters and their vehicles. "I try not to be too specific about the particular make and model of car" says Bill, "You'll always get someone saying that the spring hanger or the handbrake lever is wrong – so they're usually generic Vintage cars in the cartoon style." He's a big motoring fan, too, with a penchant for Jaguars. He runs an XK8 as his daily driver, and a "fussy" Alfa 147.

He is a real individualist and holds strong opinions on most matters. He's also a keen watcher of the world about him: that's why we all love his wry observations on our indulgent and enjoyable pastime. Here's to the next 30 years cartooning for *The Automobile*. One thing's for certain: with subject matter provided by the old-car hobby, Bill Stott is not about to run out of gags any time soon.

You can purchase or commission original artwork by Bill Stott. Visit www.billstott.co.uk

"Got that sump plug out yet, Derek?"

"I am going outside. I may be some time."

"That's right. 'Partly finished restoration project.' This headlamp's finished."

"Beautifully engineered but not much to look at. Reminds me of my cousin Doreen."

Classic, American & Veteran Tyres

From the UK's original Classic Tyre Specialists   Est. 1948

North Hants Tyres

Official UK Importer

The worlds best selling line of vintage tyres –DOT and ECE approved (where applicable). Made in original moulds to maintain authenticity. North Hants Tyres are the UK's largest stockist.

B.F.G. Founded in 1870, North Hants Tyres are the exclusive UK distributer offering many of the O/E sizes for American vehicles including the Radial T/A synonymous with 60s muscle cars and the High Pressure Silverstone for Early years.

New whitewall range for cars that require a whitewall. Over 30 sizes for your classic car. From Cadillac to R.R. Silver Cloud, we stock them all.

Excelsior Tyres are now the leading force in the Antique and Classic Car tyre market. With the recent addition of the radial tyre range made in the USA it offers the very best in quality at competitive prices. Covering Early Beaded Edge to modern Radial, Excelsior offers the complete range.

The first brand name we imported some 35 years ago, and still popular in the American car whitewall market. Competitively priced fully E-Mark ECE approved tyre.

Imported for all the Drag Race readers. All the popular sizes stocked-other sizes imported to order to guarantee new production.

For help, advice and information please call: 01252 318666

2 Ivy Road, Aldershot, GU12 4TX

Email: sales@northhantstyres.com

Web: www.northhantstyres.com

North Hants Tyres congratulate *The Automobile* on their 30th Anniversary

More than 40 years of experience specialising in pre & post war cars

Four decades of experience in the restoration, recreation and redesign of historic cars, **JSW Group** has built a worldwide reputation and the business has now progressed to three industrial units in Waterlooville with a twenty five strong work force. Consequently **JSW Group's** respect, contacts and suppliers are unrivalled in the historic and classic car world.

Restorations • Recreations • Redesigning • Engine building • Machine shop
Turn key solution • Circuit support • Storage

Work from drawings or original components • Full CNC 4-axis machining capabilities
Digitising facility • Full pattern shop association • Full foundry association

Restoration of aluminium or steel bodywork • Manufacture of complete bespoke coachwork
Chassis restoration and complete production • Wings, radiator grills, fuel tanks, etc.
Design service available • Film work fabrication

Telephone +44 (0) 2392 254 488   Fax +44 (0) 2392 254 489   Email info@jswl.co.uk   Address Pipers Wood Industrial Park, Waterlooville, Hampshire, PO7 7XU, U.K.

www.jswl.co.uk

Getting Started

Gerry Michelmore considers some of the methods – ingenious and otherwise – used for starting engines. Illustrations by **Apsley**

When I was a very young man, trying to start my car with the crank handle, an elderly gentleman standing behind me suggested I should place my thumb on the other side of the handle, rather awkwardly, alongside my fingers, if I wanted to avoid injury.

For decades, I followed his advice and passed it on to fellow motorists. It took only a few minutes of analytical thought to realise that the theory was entirely erroneous and, by dint of imparting a false sense of security, downright dangerous. I was pricked into

"OH GRAHAM, WHAT A BIG STARTING HANDLE YOU HAVE!"

action when I saw a representative of the National Motor Museum disseminating the same crackpot advice to the world through a television broadcast.

There are two principal ways to be hurt by a starting handle. The first happens when the engine kicks back, wrenches your fingers open as it whizzes anti clockwise to administer a sharp rap on the knuckles if you haven't been quick enough to get your hand out of the way. The second more serious injury occurs when you are pushing down on the handle, *ie*, as you twiddle it for several rotations; if the engine kicks back as you are pushing down, your hand is pushed upwards until it breaks your wrist. The position of the thumb makes no difference. These injuries can be avoided by having the ignition timing sufficiently retarded.

It is safe to pull up. It may not be safe to push down or twiddle. The only situation in which the thumb position is significant is when the starter dog fails to disengage and drives the crank handle clockwise. This never happens.

The propensity for an engine to inflict injury is determined by three factors: capacity of one cylinder, compression ratio and timing. A standard Austin Seven with 187cc at five to one offers no threat. I have twiddled the handle with impunity thousands of times, with no retardation. The only occasion when it bit back was when the timing had slipped. I used to gain silly pleasure from the following trick: turn the engine off outside a shop, turn ignition back on before exiting car, then, as you walk back to the driver's door, walk nonchalantly past the front of the car and, without breaking step, give a quarter-turn tweak to the handle. A warm engine in good order should always start.

On the other hand, I would not go near the starting handle of Duncan Pittaway's Beast of Turin FIAT S76 (seven litres per cylinder) unless I had personally witnessed that the ignition was distinctly after TDC.

The car responsible for this article was a 1912 Locomobile of seven litres. It was fitted with a 1913 after-market bolt-on dynastart. This 2cwt monster occupied the place of a blower on a Bentley. Only on occasion would it work. On the front of the dynastart was a dog. This accommodated three positions for the handle and looked like a three inch auger bit. On tick-over it could go through solid oak at walking pace. I was terrified that a spectator wearing a mackintosh would step back and be turned instantly into a human propeller, extremities thrashing on the road. I made a starting handle for this lethal dog using a handy Austin Seven component. It was an inadequate thing with less than half the throw of the original and made me concentrate my thinking about thumb-theory.

The relative position of the dog to the crank should be set so that compression arrives at about eight o'clock. With the crankshaft turning clockwise (anti-clockwise at the power take-off/flywheel end) this arrangement suits right-handed humans. I contend that it is our dexterity that has determined the convention of rotation. I know of only one machine that sports an anti-clockwise starting handle, the 1907 Phanomobile. The owner is not left-handed and, last time I saw him, he had a broken wrist.

When Alvis turned the engine back to front in their FWD car they were loyal to the

Getting Started

convention, but instead of the starting handle emerging from the back of the car it was made to engage with the layshaft, at some considerable mechanical disadvantage.

There is inherent safety in the starter dog engagement, so that when the engine starts, it automatically disengages. Incidentally, the dog itself is rather a difficult thing to machine but can be made simply, by using wire erosion.

A big old Gardner that lurked in the depths of Roger Hardy's fishing boat had a shaft running the length of the engine with a crank handle at each end. These handles had a massive throw, about two feet, like a fairy-tale well. The shaft was connected by chain and sprockets and a free wheel device as in bicycle practice. To start, you set the de-compressor levers on all the cylinders, took a few deep breaths and began winding. With thick cold oil, the acceleration of the two-ton flywheel was disappointingly slow, but after about half a minute or so the first one to reach heart bursting point would yell "NOW" and the nearest de-compressor was flicked on. If you had invested sufficient energy in the flywheel, it would thump into life; then you invited the other cylinders to join in, one by one. Usually, the beast wouldn't start until the oil temperature had been raised to some degree by this process. One day it thumped into life and the free-wheel device failed utterly in its duty. The appalling scene was illuminated by a meagre oil lamp. The only way to stop it was to reach past a wildly whirring crank and yank on the rack of the injector pump.

My Carden cyclecar has a cockpit-mounted kick start which is, I think, unique. Some small cars with motorcycle engines have kick starts and although it seems a perfectly legitimate way to start a motor cycle, kick starting a car is strangely undignified. With a JMB three-wheeler, the starter is on the side and when you kick it, half of your energy is dissipated in rocking the machine as if it were a boat. For a Bond Minicar, you open the bonnet, insert your right leg and kick in the manner of John Cleese administering punishment.

Some cars, like early Lanchesters, Morgans and GNs, have their starting handles inserted in their sides like keys in clockwork toys. Rather than making their arms do the work, some people seem to find it efficacious to stand on the car and jump on the handle. In the cockpit of a Trojan, you are supplied with the kind of lever familiar to signalmen, which rotates the under-seat crankshaft via a ratchet. On American high-wheelers, a Veteran version of the Land Rover, they tend to have two-litre twin-cylinder horizontally-opposed engines with their starting handles at shoulder height, conveniently placed to knock your head off when things go wrong.

Bump-starting is something we have all experienced; single-handed bump-starting is another thing. I once used a Heinkel bubble-car to get to the station and on winter mornings the battery could not be coerced into turning the engine over. The technique was to open the front door, which has the steering wheel attached via a cunning hinge, stand facing the car and pull it towards you. Then run backwards as fast as you can until you launch yourself forward, turning round, sitting down, slamming the door and knocking it into gear, all in one movement. The journey was only a mile or so and by the time we got to the bend before the station we had reached terminal velocity, nearly 53mph. One morning we arrived at this point with the driver entirely unaware that the Heinkel had a trick up its sleeve: it had sneakily drained all its brake fluid. When the pedal went, donk, to the floorboards, I heaved on the handbrake. This had the immediate and magical effect of making the landscape, trees and sky rotate in a psychedelic arc. The cabin was full of smoke and dust but I seemed to be the right way up and still breathing. Then I noticed streams of bowler-hatted gents peering at me as if I had just landed from Mars. Primarily, I felt acute embarrassment and just wanted to disappear. I thought, I wonder if it will go? I tried the starter. Amazing. I parked the car and strode into the station as if I did this every morning. As the train came in, I thought, I wonder what happened to the car? I sneaked back to look. Unbelievably, it was unmarked, except for a scrape on one side, which was absolutely vertical.

One day in November, 1974, I found myself on a train to Scotland with a bag of tools and no return ticket. With me was Tony Cox, who had bought a supercharged FWD Alvis, sight unseen. Our naive intention was to drive it home to Surrey. Our optimism was based on a 20-year-old auction catalogue

" 'E WON'T 'URT YOU, SIR, 'E JUST GETS PLAYFUL, SOMETIMES!"

"THE NOBLEST PROSPECT WHICH A SCOTCHMAN EVER SEES, IS THE HIGH ROAD THAT LEADS HIM TO ENGLAND".
DR. SAMUEL JOHNSON, 1709-84.

which described it as 'in running order'. We would go on to learn that this had been, even back then, an outright lie; it had been thrown together with components and controls fitted back to front or upside down. It had spent the intervening years in a disused gas-works half way up a mountain in Fife, where its condition had not been improved by the addition of an even layer, half an inch thick, of Scottish pigeon poo. We spent two days stuffing holes in the radiator, cleaning out fuel lines and trying to find out how it was supposed to fit together. Apart from a magneto, there was no electricity or wiring of any kind. When we tried push-starting, we were rewarded with some encouraging pops and bangs but no matter what we tried, it would run for only tantalisingly short bursts. Round and round the gas-works we pushed. Exhausted, we resolved to endure one more whisky-soaked night in the gas works, then at dawn we'd throw our things in the car and push off down the mountain. If it worked we would drive it to Kingston and if it refused, we'd come back with a trailer.

At the bottom of the mountain we had burned all our gravity and the car was popping and banging in first gear and occasionally leaping forward. Gradually more and more valves were becoming unstuck and then suddenly, like the sun breaking through, it accelerated away and went roaring down the road. With the throttle open, the engine ran sweetly, but with anything less it would promptly stop. Refuelling would have to be done at a downhill petrol station. Waiting at the toll for the Forth Bridge, the engine died. There was a ramp nearby that led up to a private car park. Our plan was to push it up the ramp and effect a restart by whizzing back down. As we were turning round, we noticed eight floors of windows lined with spectators of our enterprise. The urgency to get started was much enhanced when we spotted the big blue sign saying 'Police Headquarters, Edinburgh'.

We had taken the precaution of booking an MoT in Kingston for the following day but we were not keen to discuss our exact legal status. We made erratic progress down the A1 throughout the day, stopping at downhill repair shops to get things welded back on. We were still hundreds of miles from home as the daylight faded. We bought some bicycle lamps and tied one on each corner. The car ran on through the moonless night at a minimum speed of 70mph, rain sheeting down in torrents. The driver was equipped with a seat as well as an aero-screen. The mechanic's comfort consisted solely of a wet bag, perched on the chassis. Forward visibility was reduced to about six feet. Navigation was effected by the mechanic peering over the side and indicating to the driver by pulling or pushing on his shoulder whether we should be closer, or further from the kerb. This worked well until we hit roadworks at Watford. Literally.

When we succeeded in fighting our way through London's rush hour to reach home, we felt like heroes returning from the North Pole. Many years later, driving home from competing at Prescott in the same car, the dynamo was on strike and the fuel tank was nearly empty. It was a lovely summer evening and I could see adequately with sidelights, only using headlights to reassure passing cars. I wasn't worried about getting home because I had a two gallon can in the boot. You would think the experience on the A1 would have been seared into my mind sufficiently to warn me that the re-fuelling should be done downhill. I'm afraid the truth didn't dawn until just after the engine stopped on a perfectly flat stretch of road. Solo bump starting this car requires special athleticism, as there is but one door on the passenger side. I remembered my Heinkel days and succeeded on the first attempt.

While living in a grass hut with a beautiful, if fiery, Spanish lady, on the remote North East coast of the Dominican Republic, I discovered a VW Kübelwagen in the jungle. It had been used to death as a hire car, then stripped and abandoned. Finding a replacement clutch involved hitching a lift on

VERNACULAR ARCHITECTURE Nº 73 - DOMINICAN REPUBLIC.

Getting Started

"NATIVES HEAP RESTLESS".

a light aircraft. In the course of this flight I witnessed one of those scenes where you kick yourself for not having a camera. The sun shone on a rain shower, making a complete circular rainbow below. In the middle of this circle flew a vast flock of flamingoes over a glittering lake.

In order to take the engine in and out of a VW, it is necessary to raise the back of the car several feet. I had no jack and it was no mean task using levers, wedges and stones. With the new clutch in place, I wanted to test everything before bringing it down from its perch atop a pyramid of stones. I was thinking of micro-lights with VW engines, so I selected top gear, blocked one back wheel, got hold of the other and yanked it round. Chuff, chufff, chchufferty chuff and it was away. It made the perfect vehicle for that roadless paradise.

My two-stroke Carden engine will reverse itself willy-nilly at times of its own choosing, much to the entertainment of witnesses. A Stuart Turner marine twin has the same characteristic. A friend of mine in his Old Gaffer approached the harbour wall at a moderate pace, the bowsprit pointing the way, with the gear lever in neutral. The intention was to wait until the last moment before engaging astern and opening the throttle, whereby the way would be taken off and the stern tucked in by paddle-wheel effect in one neat manoeuvre. Unbeknownst to the proud owner, the engine had sneakily reversed itself while ticking over in neutral. Hundreds were gathered to witness the calamity. Their laughter rang round the bars for weeks.

The Messerschmitt bubble car embraced the propensity of two-strokes to reverse themselves by fitting a starter motor that worked in either direction. This gave the tiny three-wheeler the dubious advantage of having four speeds in reverse. Some larger marine engines were started by introducing compressed air into the cylinders at the right moment. You could reverse the direction by changing the air valve timing. On four-strokes, you also moved the camshaft to select astern timing. The air reservoir might hold enough air for two or perhaps three changes of direction. This forced the skipper to choose his moments to telegraph the changes down to the engine room with the utmost diligence.

Within the confines of a narrowboat, the 1500cc single-cylinder Bolinder needs to have its hot bulb scorched with a blowlamp for ten minutes before the owner takes courage, folds down the starter peg (like a theatre seat) from the face of the flywheel and tries to kick it into life. The kick is administered against the direction of rotation, bouncing the piston off compression in the hope that enough inertia is generated to carry it over compression in the other direction. The engine speed is regulated via a governor-controlled 'hit and miss' inlet valve. Wherever it voyages, it spreads a bow-wave of happiness with its uneven beat: TToom, tum, tum, tum... TToom, tum, tum, tum... TToom, tum, tum. It merrily proclaims the same message in a stream of smoke rings.

Diesel engines like to have glow plugs in their combustion chambers to help them get going. The ingeniously counterbalanced Sabb engine, designed to run underwater, eschews electricity and uses a sort of firework for this purpose. It looks like an imitation cigarette which fits into a plug that you screw in before trying to start up, and catches fire as the piston comes on to compression.

Some aircraft and tractors use the energy from an exploding shotgun cartridge as a means of starting. This combination of pyrotechnics and mechanics seems absolutely sublime. I cannot think why cars have not adopted this method. What fun to be had in the Waitrose car park!

When we have worn ourselves into a state of exhaustion and exasperation cranking and pushing, we resort to towing. It is only then that we realise we should have done this in the first place. For those in doubt, the following technique will be found efficacious. Both tower and towee should be experienced, competent and have some gumption. They should co-operate so that the tow rope stays tight at all times. The towee turns on ignition, engages second gear (third if the tow car is really feeble) then, with left foot on clutch, ball of right foot on brake and heel of right foot on accelerator, he signals the tower when the rope is tight and ready to go. The tow rope length should be one to one and a half car lengths. At 10mph the towee lets in the

"DO YOU THINK WE'LL APPEAR IN THE NEW YEAR'S HONOURS LIST?"

clutch. If the engine fires, he stands by to dip the clutch, blip throttle, apply brake and signal the tower. Simultaneously. A working handbrake and horn make useful additions.

Apart from competition cars in the paddock, you don't see it much these days. It used to be commonplace. Whenever someone hauled a wreck out of a hedge or collapsed shed, his friends would gather round for the ritual tow start. At the end of a pub meeting, it was normal for one of the habitués to have a dud battery and questionable compression. There is a particular piquancy in tow starting

"GIVE ME A BARRING ENGINE, MR. BOND, AND I CAN MOVE THE WORLD."

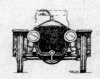

Getting Started

a very grand Hispano or Phantom with a scruffy Austin Seven. It hardly needs mentioning that there are many and wonderful permutations for things to go wrong, and for coachwork as well as egos to be badly dented.

We have all heard stories of Silver Ghosts, after slumbering for months in their motor houses, swinging themselves miraculously into life when their ignition levers are advanced. I have seen it done within a few seconds of shutting down, but that was with coil ignition. For it to be possible, you have to retard the ignition before shutting down and the crankshaft must come to rest exactly in the right position. I suspect most of these tales are apocryphal.

Some big steam engines need their crankshafts to be in exactly the right place for them to start. For this purpose holes are drilled in the circumference of the flywheel so that bars can be inserted to haul the crankshaft round. Really big engines have what is called a barring engine to do this work. You can witness the phenomenon at Kew or Kempton Steam Museums. A V-twin steam engine, the size of a Harley, drives a 12-inch pinion that engages with a ring gear some 30 feet in diameter. With this little engine chuffing away, the crankshaft is very slowly brought into the right position when the giant main engine silently accelerates into its stride. To prevent the little pinion being spun to impossible revs, the whole barring engine is mounted on a pivot so that, on starting, the main engine disdainfully shoves it out of engagement.

Very few cars adopt the non-recoiling pull-cord. One of them is that pinnacle of auto engineering, the Bédélia. You wrap a cord, or perhaps a long Isadora Duncan-esque silk scarf, around one of the drive pulleys, having moved aside the droopy drive belt and, with a grand flourish and a shout of "*Voilà!*", you haul the engine into life. Most people of my generation were acquainted with the miseries of the non-recoiling pull-cord with the Seagull outboard motor. One of the deepest pleasures, now lost forever, was watching a doomed sailor drifting downtide, beside himself with rage, as he pulled repeatedly at his recalcitrant Seagull, a stream of filthy expletives fading in the wind.

The AV Monocar uses a long spring and a chain which takes half a turn round a crankshaft sprocket and ends in a pull handle. The AV owner needs more determination than almost any other motorist. A backfire hauls him headfirst into the hot engine.

The most entertaining starting procedure I have witnessed was by the late Hon Douglas Fitzpatrick on his Maybach Métallurgique (4.5 litres by six cylinders). The driver sits grandly behind the wheel. The mechanic anoints the priming cups and inlets with generous sloshes of fuel, then with authoritative, reciprocal commands of "Switches Off !" and "Suck In!", the crankshaft is rotated twice. This is achieved by the mechanic inserting a four feet long bar in the nose of the crankshaft and turning it with great heaves, repositioning after each half turn as it reaches the ground. The engine is then set so that number six is just after TDC. The mechanic stows the bar under the running-board and takes his seat beside the driver. Unbeknownst to the on-lookers, number six plug is connected to a trembler coil, isolated by an imposing and ornate brass switch on the dash. The driver reaches forward for the switch and hesitates... for effect, as the crowd sucks in and holds it. Then, with a ground-bouncing BOOM!, the machine explodes into life and the audience bursts out laughing.

Perhaps it is worth mentioning, children, that none of this, apart from the thumb bit, should be attempted at home.

1907 AUTOMOBILE COMMENCEMENT EQUIPAGE OR "THE YOUNG MASTER PROPOSES TO VISIT HIS CLUB"

Subscribe to your favourite magazine

SAVE £29!

We realise that, for some of you, finding a newsagent who stocks *The Automobile* regularly can be difficult. This is because some of them, in addition to taking a huge proportion of the cover price, try to charge us just to put the magazine on their shelves — and we won't play ball. The obvious answer is to take out a subscription. That way, you save money and have hassle-free delivery. Better still, why not treat yourself to a two year subscription and get two extra issues free of charge? The saving then works out at a whopping £29! Just fill in the form below and post, email, fax or telephone Jules, our friendly Subscriptions Manager, or visit our website, www.theautomobile.co.uk

UK rates
- ☐ 6 issues £25 (save £2) £ ....
- ☐ 12 issues £47 (save £7) £ ....
- ☐ 24 issues £88 (save £29) £ ....
- ☐ I am taking out a two year subscription and claiming 2 free issues

- ☐ 1 Binder £9   ☐ 2 Binders £16 £ ....
- ☐ 3 Binders £23 ☐ 4 Binders £28 £ ....
- ☐ Back issues £6 each .... £ ....

Continental Europe including Eire
- ☐ 12 issues £55   ☐ 24 issues £105 £ ....
- ☐ I am taking out a two year subscription and claiming 2 free issues

- ☐ 1 Binder £12   ☐ 2 Binders £21 £ ....
- ☐ 3 Binders £30 ☐ 4 Binders £36 £ ....
- ☐ Back issues £8 each .... £ ....

Rest of the world
- ☐ 12 issues £72   ☐ 24 issues £125 £ ....
- ☐ I am taking out a two year subscription and claiming 2 free issues

- ☐ 1 Binder £14   ☐ 2 Binders £25 £ ....
- ☐ 3 Binders £34 ☐ 4 Binders £42 £ ....
- ☐ Back issues £9 each .... £ ....

TOTAL ENCLOSED: £ ....

SUBSCRIPTION / BACK ISSUES / BINDERS ORDER FORM

Photocopy this form, fill it in and post, fax or email it to the address below

NB - This binder order form is for the new size. For the old size please enquire.

My name and address

....
....
....
....
....

Postcode: ....
Telephone: ....
Email: ....

I enclose sterling cheque for £ ....
payable to: Enthusiast Publishing Ltd
OR please charge my card no:

....

Valid from....../......Expiry....../......Sec No........

Post to: Enthusiast Publishing Ltd, PO Box 153, CRANLEIGH GU6 8ZL
www.theautomobile.co.uk Email: jules.clifton@theautomobile.co.uk
TELEPHONE 01483 268802 FAX 01483 268993

For details of back issues and other reader services, visit us at www.theautomobile.co.uk

www.longstone.com   Tel:+44(0)1302 711123

R M FOWLER LTD
TIMING & DRIVE CHAINS
website : www.chain-drive.co.uk

Roller Chains, Inverted Tooth Silent Chains, Block Chains and Cycle Chains supplied for early vehicles

CONGRATULATIONS ON YOUR 30TH ANNIVERSARY

Unit 8, The Grove, Parkgate Industrial Estate, Knutsford, Cheshire, WA16 8XP, Tel: +44 (0)1565 651051

The Prop Half

Power boating evolved in parallel with motor racing in the early years. As well as being surprisingly influential, it involved many of the same protagonists. **Hiram P Matelot** tells the story

The Prop Half

The shape of high speed boats informed the design of performance cars from the start – unsurprisingly, since knowledge of flow dynamics was based on studies in water rather than air. The empirically streamlined racers of Léon Bollée, Serpollet, Mors and others cleaved the air with distinctly nautical prows, and one of the fastest machines of its era was bodied by a boat: the Stanley brothers' *Rocket* LSR car of 1905, which reputedly wore an upturned Robertson competition canoe of the type they had towed round Massachusetts on a trailer hitched to a spring scale to measure drag. No wonder, then, that coachwork design of the first quarter of the century borrowed so heavily from the maritime, from the first torpedoes, skiffs and 'submarines' to the pert stern of a Wensum and such jaunty embellishments as teak decking, portholes and scuttle ventilators.

The connection went beyond shape, however. Given the finite number of clients wealthy enough to afford their products, the burgeoning number of car makers in the early years, especially in France, competed in a febrile market in which sporting events provided a good way of generating publicity. Costly, though – the great town-to-town epics demanded substantial logistic support, as did the Gordon Bennett and other international races that followed. Although hundreds of thousands of people turned out over great distances to watch, few of them could realistically hope to buy a machine costing several times the average annual family income. The prospect of taking part in equally publicised events at a compact, glitzy venue already favoured by Europe's social élite was unmissable, which is why the international Monaco powerboat meetings of the Edwardian era were so successful. Influential, too: the demand for high outputs from lightweight power units for boats (and later, of course, aeroplanes) inevitably drove developments in car engine design.

Motor boating was nearly 25 years old when the first Monaco meeting was held in 1904. Etienne Lenoir had conducted trials of a small craft powered by his primitive two-stroke coal gas engine along the Seine in 1870; between 1882 and 1884, he installed a new four-stroke unit based on Beau de Rochas principles to average nearly 9mph along the canal from Le Havre to Tancarville. The earnestly self-aggrandising Fernand Forest, who considered himself the James Watt of petrol combustion, essayed the Forest 6hp tram engine in his iron-hulled *Volapuck* in 1885, and Gottlieb Daimler was testing one of his first 1hp four-stroke engines in a launch soon after. Successors of this machine were imported to the UK in 1891 by Frederick Simms, who gave their fuel a new name: 'petrol'. Races were held between Forest and Gallice-powered boats at Harfleur and Nice in the 1890s, but their lack of spectacle disappointed a public accustomed to grandiose sailing regattas. Nevertheless, the Hélice Club de France was founded in 1896 to organise the fledgling sport, and a race was run on a 12-kilometre course along the Seine from Argenteuil as part of the 1900 Exposition Universelle. The victor was Marius Dubonnet, head of the eponymous vermouth firm and father of Emile and André – a trio who made an indelible mark on the history of lighter- and heavier-than-air aviation in France, as well as powerboats and, most notably, the motor car.

Emil Jellinek, then the Austrian consul in Nice, entered the fray with the *Mercédès*, a 10-metre launch built in Billancourt by Maurice Chevreux and propelled by the company's groundbreaking new 35hp engine; at the 1901 Nice regatta, it secured the motor boat speed record at 20mph. Although the English-speaking world was also experimenting with high speed motor boating, France led the way in the early years. 'France gave us the bicycle, the automobile and the air machine, so it was but natural that the Frenchmen, the discoverers and the perfectors of the racing automobile, should be premier in transferring the venue from land to water,' conceded *The Motor Boat* in a retrospective piece in 1906. 'The French paid little or no attention to the motor-driven boat until after the automobile had completely captured their fancy and dominated their fashions. Until last year the English – unbelievers in any maritime affair emanating from France – refused to take the new departure seriously. To your true Britisher, any vessel which could not hold its own in the Channel or the North Sea was a toy.'

Toys or not, the precociously gifted Alphonse Tellier soon emerged as the designer to beat. His father's boatbuilding business operated from Quai de la Rapée by the Ile de la Cité, within walking distance of the workshop of Fernand Forest, whom he often visited, along with two other highly inventive minds: the aeronautical pioneer Victor Tatin, and Léon Levavasseur, the architect of Antoinette motor boats, engines, aeroplanes and automobiles. The originality of young Tellier's approach had been evident ever since he put a de Dion single in a hull of his own design in 1898, and he was among the first to embrace the hydrodynamic benefits of the stepped hull. In 1902, Tellier's 15-metre *Lutèce* boasted a pair of Panhard 40hp four-cylinder engines linked via a magnetic coupling and differential speed gear driving a three-bladed 700mm propeller. At only 600kg, the lightweight hull consisted of three layers of very thin cedar strips mounted diagonally to each other and fixed by closely spaced copper rivets (in the manner of Henri Labourdette's celebrated 'Skiff' body of 10 years later), with a canvas membrane in between each layer. Although it exceeded 18 knots, the power/weight ratio of the twin engines was a limiting factor. They were duly replaced by a single Paris-Madrid type 70hp racing unit, and in this form the Lutèce averaged 20mph in a 100-kilometre race along the Seine in 1903.

Just as in motor racing, whether a surfeit of litres trumped lightweight agility was a moot point. Tellier tried both. In 1904, his eight-metre *La Rapée II* proved almost as fast as the *Lutèce* with a Panhard four of only 24hp, but in the same year he also constructed a leviathan: *Le Dubonnet*, whose Titan engine boasted a swept volume of no less than 85 litres. A very large Dubonnet indeed, then. Fernandel's "Do 'ave a Dubonnet" catchphrase might have been coined with Tellier in mind, for he had two: his childhood friend Emile as occasional helmsman (and, later, pilot of Tellier aeroplanes), and Marius, whose chequebook matched his gargantuan ambitions and funded the development of the most powerful racing boat of its era. As Sébastien Faurès mentioned in his recent analysis of Ernest Henry in these pages, the Titan was the first dohc engine. It was described by *Power Boat News*: 'Nothing gives a better idea of modern French auto-marine engineering than the special motor built by Messrs Desmarais & Morane, makers of the Delahaye automobile. The four cylinders, with 300mm diameter by 220mm stroke, are of a special grade of nickel gun

Seen here at Monaco in 1905, the nickel steel hull of the 13-litre *Napier II* was built by Yarrow & Co at Selwyn Edge's instigation, and set the world water speed record for a mile at almost 30 knots

1. The super-slippery Stanley *Rocket* in which Fred Marriott achieved 127.7mph at Ormond Beach in 1906 was bodied by the Robertson Canoe Company of Newton, Massachusetts

2. The nautical influence is unmistakeable on the 80hp Mors which C S Rolls drove at a record 93mph at Phoenix Park in 1903

3. Tellier's lightweight Panhard-engined *Rapée III*, which proved so competitive at the inaugural Monaco meeting in 1904

4. This rakish Grégoire torpedo was bodied by Alin & Liautard of Courbevoie, who also built a sensational six-litre Théophile Schneider skiff, complete with stern post

steel, bored from solid with very thin walls, the material alone costing $560, while the cost of working them is about the same. The jackets are of drawn copper spun with corrugations to allow expansion and contraction; a flange is turned inward on the upper end to form a gasket between the top of the cylinder and the head, which is bolted on by studs. The heads are of cast steel, each with two large oval openings, over which are bolted spiral elbow castings forming the valve chambers and pipe connections for inlet and exhaust. The combustion chambers are designed in globular form with no breaks or obstructions to the easy flow of gases. Each casting contains a nest of three valves, with external springs. There are two half-time shafts, lying along the heads in bearings machined in each elbow casting; from the upper part of each bearing project two lugs bored for the pins of the rocker shafts which operate the valves. A steel tube carries a gear drive to the pair of half-time shafts whose cams operate the rocker shafts and valves. There are three inlet and three exhaust valves per cylinder. The crankshaft is of the finest grade of steel and is hollow; its cost is $300. The connecting rods are of nickel-steel, hollow and trussed; the lower end of each forms a very long bearing on the crank pin, but the cap is in the form of two small brasses held in place by a U-bolt and nuts, leaving the crank-pin open for lubrication. A double ignition system was planned with the idea that perfect ignition of so large a charge would not be possible by the usual method at extreme speeds, but this proved unnecessary. Both magneto and battery, with jump-spark, are installed, but each alone has proven sufficient. The motor weighs 2500 pounds (three times the weight of its hull) and is expected to develop 300hp at 700rpm.' As well as winning the Prince of Monaco Cup for the fastest flying kilometre in 1905 and the next year's Championship of the Sea, a free-for-all over 200 kilometres, this beefy 24-valve machine captured the world motor boat speed record at 33.8mph. It was difficult to overtake in more ways than one. According to *The Motor Boat* in 1906, by which time the Titan had been installed in the *Delahaye*, 'great jets of flame came spluttering from her side like a 12-pounder gun, promising to be very unpleasant for any boat that chanced to get close alongside.' Fittingly, the surviving example is classed as a *monument historique*.

Across the Channel, *Daily Mail* proprietor Alfred Harmsworth had in 1903 endowed the British International Trophy to be contested during the week of the Gordon Bennett Cup, for vessels up to 40 feet and with any type of motive power. The first event, in Cork's Queenstown harbour, was won at an average of nearly 25mph by the magnificent Dorothy Levitt, self-confessed 'motorina and scorcher', in Selwyn Edge's *Napier I*: a flat-bottomed 40 foot steel 'skimmer' hull powered by the same engine as the firm's racing car, producing 75bhp at 800rpm.

But it was Monaco that stole the limelight. Founded in 1904 by the principality's Societé des Bains de Mer and Georges Prade of *Le Sport*, the first Motonautique Internationale de Monaco offered 100,000 francs' worth of prizes and the opportunity for motor manufacturers to showcase their latest technologies to those who could afford them – and they invested accordingly. The event was soon attracting dozens of the fastest boats in Europe. 'The peculiarly advantageous conditions which obtain at Monaco are possible nowhere else, and its situation within easy distance of all the Continental motor factories and boat shops, its ideal scenic environment, its date in the first warm days of Spring and the liberal prizes almost equalling the value of the boat, have all helped to make it easily the greatest event of each year,' concluded *The Motor Boat*. It was staged over four days, the first three for testing before competition began, with an accompanying exhibition of the latest vessels. Entries were divided into racers and cruisers (which had more decking) in four classes: less than 6.5m, 8m, 12m and 18m. The polygonal course varied from year to year between 6.25 and 12.5km.

Most of the early factory entries consisted of engines from that season's Grand Prix cars in suitable lightweight hulls. The winner of the inaugural 200-kilometre race was Léon Théry in the *Trèfle-à-Quatre*, which Henri Brasier used to test the 84hp unit destined for Georges Richard's Gordon Bennett entry that Théry also drove. Much the same was true of the Mercedes, Grégoire, Hotchkiss, Napier, Berliet, Mors, Darracq and de Dietrich entries. However, the fact that only three of the racers entered in the 150-kilometre race crossed the finishing line suggested that dedicated marine power units would be more competitive; car engines, however advanced, could not always withstand the sustained extreme conditions to which they were subject. The surpassingly elegant *La Parisienne*, for example, its steel hull crafted by the Levallois coachbuilder William Lamplugh, was powered by a trio of Mors racing engines with a total of 90hp, but the fuel pipe unions sprang leaks with predictably fiery results.

Tellier's new lightweight Panhard-engined *Rapée III* showed the way forward, trouncing all comers in the 150-kilometre event and finishing almost an hour ahead of the second place boat, the *Princess Elizabeth*, whose Delahaye unit was of the same 7.4-litre capacity. This victory was partly due to Tellier's clever, hydrodynamically clean hull, with its greatest cross section aft and an almost circular midships flattening rapidly towards the stern, and partly because of factory-applied modifications he had specified for his Panhard engine. Weighing only 5kg per horsepower, this 40hp unit could sustain 22 knots, running at almost 1000rpm for hours. *The Engineering Magazine* approved: 'Speeds of rotation have been considered enormous, but piston speeds are lower than those of the torpedo-boat engines or express locomotives, and the weight of the reciprocating parts is much lighter. Mr Tellier maintains that such motors give no trouble, and cites the fact that in the races at Monaco, the engines of *Rapée III* were operated for six consecutive hours at 950rpm, or a total of 342,000 revolutions, and 684,000 explosions. The speed over the 16 laps did not vary more than three seconds per lap, showing the uniformity of the motive power under these trying conditions.'

The world's first true offshore event was held between Calais and Dover later that summer. Never one to miss an opportunity for free column inches, Selwyn Edge entered *Napier Minor* to represent England, running against the *Trèfle-à-Quatre*, which was the only one of the 21 starters not to reach Dover harbour. Humiliatingly for the domestic press, the Napier was narrowly beaten by *Mercedes IV*, which made the choppy crossing in one hour flat.

For the second Harmsworth Cup at Cowes in September, Edge entered *Napier II*, built by Yarrow of Poplar and powered by a brace of 45hp Napier racing engines with synchronised high-tension ignition and exhausted via funnels, both engines being connected by Napier metal-to-metal marine clutches running in oil. The Napier's opponents included Adolphe Clément's *Bayard*, the *Trèfle-à-Quatre* and a lone US entry, Smith & Mabley's *Challenger*. Makers of Simplex and, later, Crane-Simplex cars, Smith & Mabley was one of Manhattan's first imported car dealerships, distributing Panhards, Peugeots and Renaults. Their first motor boat, the *Vingt-et-Un* (after its speed in mph) was designed by Clinton Crane for a Panhard four-cylinder with the Krebs automatic carburetter, mounted behind the helmsman at his inclined steering wheel. Like the subsequent *Vingt-et-Un II*, *Challenger* had Simplex's own 150hp straight-eight, but, according to *Scientific American*, 'the ineptness of the mechanic in failing to clear minor ignition trouble' meant that England's Napiers retained the trophy at a speed considerably less than that of which *Challenger* was capable.

Napier II and Lord De Walden's *Napier* were among more than 100 entries for the 1905 Monaco. Gobron-Brillié installed one of their 100hp opposed-piston engines in a hull by Gustave Pitre. He had built the victorious cross-Channel *Mercedes IV*, which was also competing, along with the *Mercedes-Mercedes* and the twin-engined *Mercedes Charley-Pitre* (Charley being the Mercedes agent in France). Renault also used a Pitre hull for the *Billancourt*, with an updated Paris-Madrid powerplant, modified with lugs on the aluminium crankcase for attaching to the hull. The cylinders cast in pairs provided a common exhaust chamber with separate inlet chambers on each side of it and a very light crankshaft of high grade nickel steel with a single central bearing, but it required the service of an extra man as the engine couldn't be operated by the helmsman. *The Suzon-Legru*, with twin four-cylinder Hotchkiss units developing 170bhp, proved fast. Turcat-Méry entered the twin-engined *Pi-ouitt IV*, Brasier now fielded the *Grand-Trèfle* and Mors the *Force Pas*. De Dietrich, Berliet, Panhard-Levassor, CGV, Delahaye and FIAT entered eponymous boats. Maurice Farman and Alfred Neubauer entered two fast Tellier-designed craft named after their retail emporium, the Palais de L'Automobile: the 12-metre *Palaisoto I*, with a triple-carburetter crossflow dual ignition 23-litre Panhard six-cylinder which had exceeded 27 knots in trials on the Seine, and the smaller eight-metre *Palaisoto II*, with a Renault four-cylinder from the previous year's Vanderbilt Cup car. Another Tellier hull, *La Turquoise*, was conducted by the indomitable sportswoman Camille du Gast, and Tellier himself competed with the most lauded boat of 1905 and successor to the *La Rapée* series, the super-light Panhard-powered *La Rapière*, which finished second to the works Panhard entry after *Napier II* gave up the ghost on the final lap. Within weeks of the event, Tellier was hired to take part in pivotal and internationally publicised trials to determine the tractive effort required for powered flight by using *La Rapière* to tow the cellular gliders of Gabriel Voisin and Louis Blériot along the Seine.

Another contender in the eight-metre class was massively influential in the wider engineering context: Léon Levavasseur's *Antoinette*, powered by a marine version of the innovative lightweight V-8 that gave Europe wings. 50hp models of this engine powered Alberto Santos-Dumont's gawky *14-bis* for the world's first officially observed flight in 1906, Blériot's first tractor monoplane and Paul Cornu's helicopter (the first to lift a man off the ground) in 1907, and Henry Farman's Voisin for the world's first closed circuit kilometre and town-to-town flight in 1908.

An electrical engineer, Levavasseur determined to build his own i/c engine after seeing Daimler demonstrate his motorboat on the Seine in 1887, and started work in 1901 on a unit that would yield one horsepower per kilogram. With finance from Jules Gastambide, the first engine was bench tested in Puteaux a year later. He made the mistake of prematurely fitting it to an airframe of his own design, which crashed on its first flight; he then fitted a second engine to a light eight-metre hull, which he christened *Antoinette I*, after Gastambide's daughter. Levavasseur opted for a 90deg V-8 on the basis that the inherent balance of this configuration and the number of firing strokes per revolution obviated the need for a flywheel.

The Prop Half

1. The two 40hp engines of Tellier's elegant 15-metre *Lutèce* of 1902 were replaced by a Paris-Madrid type 70hp racing unit the following year

2. A studious-looking young Alphonse Tellier, one of the most influential innovators in early European powerboating

3. Renault's Pitre-built racer, *Billancourt*, being lowered into the water in 1905

4. Beside the hull, a V-24 version of Léon Levavasseur's groundbreaking Antoinette V-8, one of the most important engines of the era

5. The handsome and effective hull of Mackay Edgar's successful *Maple Leaf IV* was built by the leading UK constructor, Saunders of Cowes

6. *Sigma IV*, with its fearsome 200hp four-cylinder Despujols, in 1913

7. The Brazilian aeronautical pioneer Alberto Santos-Dumont crouches determinedly at the helm of his strikingly innovative *No 18* hydroplane

8. With his characteristic floppy hat, Santos tends to the Antoinette V-16 improbably supported on thin struts above the pressurised fusiform floats of *No 18*

9. Racing driver Léon Théry cruises the Richard Brasier-engined *Trèfle-à-Quatre* into the bay of Monaco

10. With an air of Manuel from *Fawlty Towers*, a hapless Antoinette employee demonstrates the relative lightness of the 100hp V-16 for publicity purposes

11. A Lorne Currie design, the elegant Mors-engined *La Parisienne* (here on the slipway at Monaco in 1904) was built in steel by Lamplugh of Courbevoie

12. *Le Dubonnet* at Monaco, with its monstrous Titan four-cylinder engine

Its flat plane five-bearing crankshaft ran in a webbed aluminium crankcase with each cylinder enveloped in a separate water jacket – initially of spun brass, and then of copper, electrolytically deposited by a lost wax process. The steel cylinders themselves were machined to a thickness of little more than a millimetre. Two generously sized valves per cylinder were located one above the other at the side of each hemispherical combustion chamber, the inlets automatic and the exhausts actuated by a central camshaft. Levavasseur's ingenious direct delivery fuel system did away with carburetters and induction pipes: individual plunger pumps operated by variable throw eccentrics fed fuel directly into separate inlet ports where it was vaporised by the aspirated air, the flow being regulated by altering the travel of the eccentrics. Antoinettes could be run in either direction, via a small handle at the end of the camshaft that rotated the cams through 90 degrees. These light, compact engines had flanges at both ends of the crankshaft to allow two or more to be mounted in series, so 16-, 24- and even 32-cylinder variants made their appearance (although the latter sank the hull into which it was fitted).

The Antoinette's power-to-weight ratio was unrivalled. Before long, boats of the same name had secured world records for every distance from one to 150 miles. At Lake Garda in 1906, *Antoinette III* (with a 220hp, 42-litre V-16 weighing 340kg installed in the hull of the former *Billancourt*) inherited the epoch-defining mantle of *La Rapière* by averaging more than 30mph over 150 kilometres. The Brazilian pioneer Alberto Santos-Dumont used a smaller 128hp version of the V-16 for his 250kg hydroplane *No 18* the following year, with the intention of exceeding 100km/h (and thereby winning a 50,000-franc wager with Charron of CGV). Improbably suspended on six slender struts high above the central fusiform coque of rubberised canvas inflated by compressed air and driving a large three-bladed airscrew, the Antoinette looked every inch the elegantly minimalist engineering masterpiece, but Santos lost his bet.

The following year, 1907, saw the emergence of a new class at Monaco, for hydroplanes. Paul Bonnemaison, who had patented a single-step hull, fielded the tiny but effective *Ricochet-Nautilus*, with an air-cooled 10hp Buchet triple. Levavasseur's first stepped hull was the *Obus-Nautilus*, consisting of two connected floats, one carrying the engine, the other the pilot, controls and fuel tank. It won the new class, and such hulls proved so successful that by 1909 all classes, racers and cruisers alike, had entrants so equipped. Admitting that 'most of the spectators had not the faintest idea as to the theoretical reason for their shape,' *The Rudder* described one such 'interesting type of marine freak' thus: 'A single hull with an abrupt vertical jog in the bottom about midships, with the exception of one which is a series of athwartship boxes held together by girders and driven by an aerial propeller. Afloat, the craft resembles a windmill mounted on horse troughs.' This ungenerous description referred to Count Charles de Lambert's entry, in which a 70hp Antoinette engine drove the airscrew; unlike Santos's *No 18*, it aroused great interest by its extraordinary speed before immolating itself. Hydroplanes per se were not new; the doyen of French aeronautical pioneers, Clément Ader, had built models of stepped hydroplanes in 1895 – he even proposed injecting compressed air underneath the surface of the planes to minimise resistance. De Lambert had begun experimenting with such hulls even earlier, in 1892, well before the availability of engines sufficiently light and powerful to take full advantage of the skimming principle. He had tried his first 'gliding boat' on the Thames in 1897 – a catamaran of two long, pointed floats between which he placed adjustable flat plates that lay below the water, powered by a Field steam engine, too heavy for the device to ride up upon them. He had more success with a de Dion single in 1904; by 1913, his fellow Wright pilot Paul Tissandier had taken the de Lambert *glisseur* to within a whisker of Santos's 100km/h.

Levavasseur's second hydroplane was more conventional. 'In front it carries a flat-bottomed boat of the usual form, while in the rear is attached a float forming a tailpiece which aids the gliding action. The front boat carries the motor and passengers. From the motor a shaft runs to a propeller carried in the rear of the tailpiece, where the under-surface of a box-shaped piece forms a plane, and the whole device gives a constant angle in order to secure the gliding action on the hydroplane principle in smooth water,' explained *Scientific American*. 'The 50hp motor weighs 133 pounds, and showed a good speed upon the Seine. Santos-Dumont piloted it on different occasions and was favourably impressed with its performance, stating that it had a good balance and gave almost the same sensation as an aeroplane.'

The comparison is significant because, although powerboat development continued unabated from 1908 onwards (especially in the small hydroplane class, such as the *Ricochet* series), public attention shifted to the novel aeronautical goings-on at Issy-les-Moulineaux and at the first great air meetings throughout France, which starred many of the same protagonists. Tellier began making his own monoplanes and seaplanes for the Dubonnet brothers to fly, and Levavasseur concentrated on building his supremely elegant Antoinette monoplanes. Panhard, Lorraine-Dietrich, Renault, Clément Bayard, Gobron-Brillié, Mercedes, Grégoire, FIAT and many others turned their attention to designing aero engines. There were exceptions, of course – sometimes, surprising ones.

In 1910, for example, Herbert Austin built an all-aluminium 300hp marine engine, said to be desmodromic, for the *Irene I*. Its second 380hp incarnation briefly captured the world water speed record on Southampton Water. The firm also played a big part in the success of *Maple Leaf IV*. Powered by two V-12s designed by the New Orleans Company, its aquaplane hull was built by Saunders of Cowes with five steps of varying cross section, rising to the third at maximum speed. The owner Mackay Edgar asked Austin to improve the engines at Longbridge, where new cylinders were cast with larger coolant passages, the pistons were lightened, the valve train redesigned and a new pressure-fed lubrication system installed. An ABC flat-twin was used for starting. With Tom Sopwith at the helm, it defeated all comers at the 1912-13 Harmsworth Trophy in the US, and was the first boat to average 50 knots.

By then, though, such adventures by mainstream car makers were isolated rather than commonplace. Although the glory days of powerboat racing were yet to come, and the links between automotive and marine sports continue in some form today, the era of parallel development and technology transfer was largely over.

Thanks to Sandy Skinner and Francis Metzger for their help in compiling this article.

FVR FORMHALLS Vintage & Racing Ltd.

enquiries@formhalls.com **01725 511684** www.formhalls.com

Formhalls Engines

Aircraft quality & reliability for competition & road cars

- Full or part engine rebuilds to concourse standard with photographic build sheet for your car log

- All engine machining & balancing

- Formhalls - Hoyt white metal bearings
  Fully machined & line bored, etc. or rough cast for you to finish
  Guaranteed fault free for the life of the engine

- Mechanical rebuilds inc. transmission, steering and suspension etc.

- Collection & delivery service

Parkers Close, Downton Business Centre, Downton, Wiltshire. SP5 3RB

FE and FENC: the Isotta Fraschini voiturettes

Pages 64-65. Alfieri Maserati about to take the start at Dieppe in 1908 (*The Grand Prix Library*)

1. This handsome FENC had some limited weather protection but no windscreen (*Tito Anselmi Archive*)

2. The UK office of Lorraine-Dietrich issued a sales brochure in 1909 featuring an Isotta FE (*Warwick Anderson Collection*)

3. Many drawings of the FE and FENC survive, and have been invaluable in recent years when restoring and maintaining the surviving cars (*Peter Latreille Collection*)

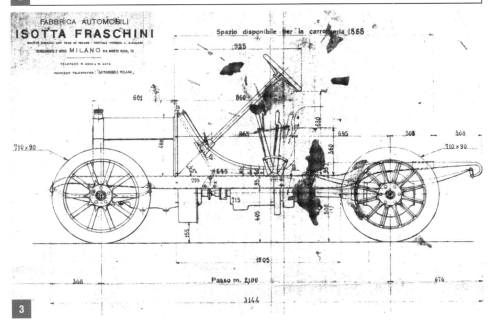

whereas the Dieppe cars had only three.

Lorraine-Dietrich had a shareholding in Isotta Fraschini at this time and were considering UK manufacture of a range of models based at the old Ariel factory in Selly Oak, Birmingham. That had been purchased by a newly formed UK subsidiary, according to *The Motor* for 14th April, 1908. Lorraine-Dietrich mounted a sales effort later in 1908 for the FENC re-badged as a Lorraine-Dietrich in the UK, as reported in *The Motor* for 28th July. A four-page sales brochure issued by Société Lorraine de Dietrich & Cie de Luneville from 5&7 Regent Street survives but the photographs are definitely of an FE, not an FENC. The detailed pictures of the engine are seemingly of a development unit since they are neither exactly FE nor FENC. I think any cars that were sold in the UK were badged as Isotta Fraschini, but I cannot be certain.

Overall, I suspect that well under 100 were built (see below), of which only four are known to survive. Before going on to describe the history of those survivors, it is worth recounting some history of other cars, both FE and FENC. Unfortunately, none of the three Type FEs seems to have have come down to us.

The USA – an FE

One Type FE went almost immediately to the USA and was driven by Al Poole in the Light Car Race at Savannah on 25th November, 1908, the day before the main race. Poole was race number 6 and finished fifth in the 196-mile race round a shortened circuit that was 9.8 miles long (a longer circuit was used by the Grand Prix machines), lapping consistently in 12-13 minutes. The car is pictured on page 20 of Julian K Quattlebaum's book *The Great Savannah Races*. Unfortunately, I cannot find any other race results for this car, although surely Poole would have competed in many more events. It would seem that this was the same car photographed in the USA and shown on page 179 of Tito Anselmi's book *Isotta Fraschini*, although the location is unknown (maybe New York?).

Of course, the car could have been on loan (Isotta Fraschini ran a number of larger-engined racers in the USA before WW1) and then moved back to Italy or elsewhere. Having said that, it was photographed at some point

with a New York registration number (75266) with kids playing in the car, so maybe it stayed in the USA as someone's runabout, especially since it had a silencer by that time. Whatever, it has disappeared.

**Argentina –
another FE and maybe an FENC**
Thanks to Guillermo Sánchez, I can confirm that another of the Type FE cars was sold to Francesco Chiesa, a rich businessman from Rosario. Chiesa later owned a number of other racing cars, including the Packard Twin Six which won the 1923 Gran Premio del ACA (Automóvil Club Argentino). He sponsored the Rosario-Santa Fe-Rosario race, a classic among the local 1920s open road races, under the Copa Chiesa name. These were run from 1918 to 1926 inclusive. The FE (race number 1) was driven by Chiesa at the 1909 Mar del Plata race organized by the Touring Club Argentino but he dropped out with spark plug trouble. The race was run on 10th March and covered 600km. The two photographs that Guillermo sent me confirm that this was definitely one of the FE cars.

That then raises questions about the well known photograph of Alfieri Maserati posing in what is captioned as a modified Type FE. The image has been used in Tito Anselmi's book, in *Maserati* by Orsini and Zagari and in Bob King's book *The Brescia Bugatti*, supposedly taken whilst Maserati was on a promotional tour in Argentina in 1911. That picture shows that the fuel and oil tanks at the rear have been re-located in somewhat similar fashion to their layout on the FENC town cars as shown in Anselmi's book on page 252 and 253, with a spare wheel added at the back. More intriguingly, there seems to be only one lever near the driver's right hand and, since there is no sign of any central lever, it must be a gear lever. Therefore the braking must be solely via a foot brake. Studying this photograph further makes me believe the subject was actually an FENC because, although the radiator seems to be in the more rearward position of the FE, there is no sign of an exhaust pipe on the driver's side – a key tell-tale difference between the two models. As far as I know, there are no survivors from Argentina.

The UK – the third FE and two FENCs
The third FE came to the UK and was fitted with road equipment. It was featured in *The Autocar* for 25th February, 1911 (page 254), under the heading Famous Cars in Retirement, when owned by a Colonel Dawson of Lowestoft. Fortunately the two photographs survive with LAT, although seriously retouched. The registration number is incomplete in these photos, ending in 487. The Brooklands Society have checked the

1. Al Poole raced one of the Dieppe cars at Savannah in 1908. This is the scrutineering photograph from the Automobile Club of America (*Georgia Historical Society*)

2. An FE raced in Argentina (*Guillermo Sánchez*)

3. Lionel Martin fitted his first engine into an FENC (*Light Car*)

4. An FENC was used as a course car at a Kop hill climb before the First World War (*A B Demaus Collection*)

5. Colonel Dawson converted one of the FEs to road use in the UK but apparently never competed in it (*LAT*)

entries and can find no record of any FE or FENC racing there. As an aside, Dawson also had a Type 10 Bugatti.

An FENC was used as a course car to formally close the public road at the Herts & Bucks MCC's Kop hill climb on 3rd May, 1913, as pictured in *The Automobile* for February, 1991. The registration number is not completely visible, but it looks to end 3890 and certainly not 7983 (see below).

More importantly, another FENC was acquired by Lionel Martin and the engine replaced by one of his own design, manufactured for him by Coventry Simplex. This was a four-cylinder side valve unit of 1389cc (66.5 by 100mm), the forerunner of the Aston-Martin line. The FENC in question was registered with plate LH 7983 and its impressive performance was written up in glowing terms in *The Light Car and Cyclecar* for 2nd November, 1914 (pages 598/9, with two photographs), as follows:

'After lunch somebody suggested that an attempt should be made on The Cindertrack Hill, which is situated about 1½ miles out of Brighton, on a by-road leading to the left of the main London road. Accordingly, several of the more enterprising members of the club set out, and led by Mr Martin flew off to the ascent. The hill is well named The Cinders, for the track is rather loose, and it gets steeper and steeper until the last 30 or 40 yards has a gradient of quite one in three. It was this that proved the Waterloo of the majority. Mr Martin, however, was not to be denied, and after re-starting on the one in three with one passenger he afterwards performed the same feat with two and three up besides himself, and considering his chassis weighs about 15 or 16cwt, this is a marvellous proof of the power of the Aston-Martin engine.'

Some people believe that this became the first Aston-Martin – 'A1' – that was raced at Brooklands with the nickname *Coalscuttle*, but Neil Murray, who specialises in the early years of Aston-Martin history, assures me that was a different chassis altogether. Certainly photographs of *Coalscuttle* from the rear show a completely different rear spring hanger arrangement and a different differential casing to the Isotta's.

The survivors

The USA

In the programme for the Sheepshead Bay board track races on 2nd October, 1915, there is an advertisement by Isotta Fraschini Motors Company of 2 West 57th Street in New York promoting the Scripps-Booth car, the engine of which was 'a Sterling, of valve in head type, this head being detachable' of 1700cc. Both Sterling and Mason engines were made by arms of William Durant's empire, which was about to merge with General Motors. It seems that the Isotta agents also offered a Scripps-Booth engine as a more powerful replacement for the original one in IF voiturettes. The first US survivor had had that transplant, presumably made at that time.

My old friend Gordon Barrett recently sent me an extract from a letter he had received back in 1994 from Smith Hempstone Oliver, who says: 'Before the war, I one day saw outside a garage or junkyard, next to the sidewalk, such a car. It was uncovered and deteriorating. I did not have my camera with me so no pictures were made. The garage was just east of the Boston Post Road, just inside the New York City limits. I have no idea about the fate of this car but think that Austie Clark once told me he knew something about it. I don't think he acquired it however. In the meantime he has died so you cannot pursue it with him. On top of all that, on 9th August, 1941, I discovered such a car in good condition but with a Scripps-Booth engine in a garage in Tuxedo Park, NY. They surely could NOT have been one and the same car.'

So two cars in the New York area in the late 1930s and early 1940s, surely the same two that survive today? A 1949 New Jersey title survives, with Scripps-Booth engine number 3005, giving the owner as Sam Braen of Wyckoff, New Jersey. There is a photograph on the local historical society website showing Braen, who was the local fire chief at the time, smoking a cigar at the wheel of the car looking pretty original apart from its power unit. On the radiator is a message to the effect that the car was used by another Fire Department. I have yet to find a good enough print to identify which one, although it is just about possible it was Tuxedo Park. The wings and headlamps look very similar to those of the FE discussed above – and Al Poole was the mechanic looking after Sam Braen's cars – but other features indicate that it was an FENC. This car was painted red (as you would expect for a Fire Department-related vehicle) and featured in the early 1950s in a staged race for a film about Henry Ford, presumably whilst in Braen's ownership.

The Braen family businesses are still active in the area and I asked Gordon Barrett if he could make some calls for me. It turns out he grew up in Wyckoff and, as a child, recalls peering through the windows of Sam Braen's garage at all the Brass Era cars in his collection – what a coincidence.

Who owned the car after Sam Braen is uncertain, but a subsequent owner, Vince Hanlon, recalled that he had acquired it from an estate in Briarcliff, New York, in 1977. Hanlon lived in Suffern, New York. I spent the summer of 1968 working for International Nickel in their laboratory outside Suffern and lived in the town for two or three months. Another coincidence.

The second FENC was obviously deteriorating when spotted by Smith Hempstone Oliver and it was later seen by Ralph Stein and Austin Clark in the famous

1. Although fitted with a Scripps Booth engine, this FENC seems to be in remarkably good condition in the early 1950s whilst in the ownership of Sam Braen (*Wyckoff Historical Society*)

2. This appears to be one of the FE cars, adapted for road use in New York (*George Wingard Collection*)

(or perhaps infamous) Mike Caruso scrap yard at Hicksville, on Long Island near New York, and described by Ralph Stein in his book *The Great Cars* as follows: 'The wooden spokes of its wheels were rotted away from years of being sunk into the ground. Its bodywork was gone except for the whitened wooden remains of its seats from which a few scraps of dry leather still fluttered. Its chassis was rusted. Its hood was gone. But sitting there, unsheltered, its aluminium oxidised white, was as pretty a little overhead-camshaft, four-cylinder engine as you'd want to see.'

Who rescued it from there is unknown but it was acquired in April, 1969, by Vince Hanlon – *ie*, some years before the previous car. I am not sure exactly how much work Hanlon did on his two cars, although he attempted to make a new body for the Caruso car. They were eventually located by Jeffrey Vogel in 1983/4 via an advertisement in *Hemmings Motor News* and Jeffrey's friend Bob Rubin bought them in 1985. He agreed that Jeffrey should oversee the restoration of both cars and they were shipped to Gigi Bonfanti's shop in Bassano La Grappa, in north-east Italy. During the long restoration, Gigi died and the work was completed with the help of Gianni Torelli. The deal between Bob and Jeffrey was that Jeffrey would pay for the restorations and keep one car, with Bob retaining the other.

It was decided that the engine in the Caruso car was too badly damaged and corroded to rescue, so two new engines were made by Gianni Torelli in Campanola. Let Jeffrey Vogel take up the story: 'It took me about eight years to have both cars redone to running, driving vehicles, during which time Bonfanti passed away. The car that passed from Bob Rubin to Dean Butler was slightly more original in the restoration details; both cars have about the same number of factory bits, basically all running gear except engine. The Rubin/Butler car had followed the factory plans very closely and had fabric joints in the drive shaft, a very primitive oiling setup, and cam timing as stock from the factory. The only part not to spec was the carburetter; we used a period Memini rather than manufacture a replica IF unit. The car had limited performance and was difficult to drive smoothly due to cam timing and fabric joints (rather like a worn-out Lotus Elan with wobbly donuts in the half shafts). I don't recall Bob ever driving it but it's a long time ago...

'The second car was completed with Gianni Torelli after Bonfanti's death. Based on the experience with the other car, we used universals in the drive shaft, modified the cam timing for modern petrol, added to the compression ratio and used an oiler pump and distribution system from a 1910 Züst. We fitted a Zenith updraught carburetter and heater pipe and I did drive this car for about five or six years; it is the one that Griff Borgeson drove in the article and I used it in a raft of prewar and hill climb events. Problem was that, while the engine was scaled down, the rest of the car wasn't; gearbox and diff are the same massive units from the larger IF and the car was just way too heavy for engine output. I had it up to about 100km/h on a flat road using smaller tyres, but any incline would slow the car and acceleration was very slow and brakes, of course, non-existent. We eventually put in modified brake shoes in the rear to get some stopping power.'

The restorations took a long time but finally the Caruso car with replacement engine, original gearbox and new body was completed and delivered to Bob Rubin around 1995, although the original crankcase ended up in Australia, along with the patterns for new castings, where I photographed it just after Christmas, 2011. Bob recalls that the car drove 'like an agricultural Brescia Bugatti', although on questioning he conceded that it 'went better than that early Brescia I bought in Paris'. He drove it a few times, including some laps at Bridgehampton racetrack (which he owned) on Long Island, before offering it at the Brooks auction in Carmel in August, 2000, where it failed to sell but subsequently passed to Dean Butler.

Dean told me the car would run fine in neutral and on the flat but then ran out of breath very quickly going up a gradient. Chris Leydon took the engine apart trying to find the problem but could not track it down. It may have been a leak in the intake manifolding (integral with the new block). He tried replacing the old Solex carburetter that it came with by a 1¼in SU, but there was no improvement.

Subsequently, the car was offered at auction by Bonhams at their sale at the Larz Anderson Museum outside Boston in October, 2008, when it was purchased by the late Tom Mittler, a great collector of classic speedboats. Sadly he died in 2010 and the car is presently for sale via Morris & Welford. Peter Latreille believes the new castings may be slightly out, thus causing the breathing problems.

Meanwhile Jeffrey Vogel was enjoying the experience of Isotta motoring in Italy once the restoration of the second car had been completed by 1997. He drove it there for a number of years and then sold it to Milanese enthusiast Giacomo Tavoletti. The car is on display at his amazing Museum of Communications to the north of Milan, but sadly he is not well at present and I have been unable to go and view the car nor study the information and photographs that Jeffrey passed to him when he bought it.

Australia

Charles Kellow owned a major car agency business at 206 Russell Street in Melbourne and a photograph survives of two Isotta FENCs in his showroom in 1909. Whether they are the two Australian survivors is unknown. The registration of cars only started in 1910 and existing vehicles were not necessarily numbered sequentially. By 1912, there were 5600 registered but the directories that are available today (which are not the governmental records) only cite the registration number and a name and brief address for each entry. There are no details of the make or model of the car. However, from photographic evidence it is certain that two FENCs were numbered 1041 and 1411.

FE and FENC: the Isotta Fraschini voiturettes

The former was registered to a Mr G E Thomas of Williamstown. The history of the latter is the more straightforward to recount so I will start with that. The owner was Augustus Wolskel, a local businessman and general manager of the Phosphate Cooperative Company of Australia in Geelong from 1919 to 1939.

Wolksel died in 1949 and the car was purchased from his estate by Lyndon Duckett. The Victoria State registration 1411 was still visible on the car, which was complete except for tyres. It was restored by Duckett and was seen at a number of historic meetings in the Melbourne area in the 1960s. After Lyndon's death, the car passed to his sister Beverly Greig who has entrusted its care and upkeep to Ian Morrison. The car had a major refurbishment at the Delage Garage in Melbourne in 2004, overseen by Ken Styles.

Ian kindly drove this car round to Peter Latreille's house in December, 2011, so that I could inspect it in some detail. After lunch we went for a drive and it pulled well up the hill round the Botanical Gardens, in contrast to Dean Butler's experience with his car. This example still has its difficult Isotta carburetter fitted, unlike all the other survivors. This means there are three driver-operated levers on the steering column, an advance and retard and two carburetter adjustments. For a small-engined car that is more than 100 years old, I was impressed by the performance although, as with any two-wheel-braked car, some anticipation of retardation requirements is needed.

The second car in Australia was discovered in 1968 by Bentley enthusiast John Cresswell on a farm about 200km out of Melbourne, although the exact location is now uncertain. The chassis had been chopped behind the gearbox and a belt-driven arrangement rigged up to drive a water pump off the flywheel, but the discarded rear portion of the frame was rescued from the farm as well. Only minimal progress was made at that time on restoring the car, although Cresswell partially rebuilt the

1. Two FENCs make a splendid sight in a Melbourne showroom in 1909, although it is not certain that these are the two survivors in Australia today (*Peter Latreille Collection*)

2. An amazing find! It is 1950 and car 6006 emerges from storage in Australia, having been purchased by Lyndon Duckett from the first owner. The car is now owned by Duckett's sister, Beverly Greig, and is maintained by Ian Morrison (*Chester McKaige*)

3. The lovely little engine seen from the inlet side. This is the Anderson/Latreille car, known as 6017. The manifolding was cast into the block/head unit. The magneto was relocated to the front on the FENC compared to the FE's, which intruded into the cockpit like later Bugattis (*Rob Imhoff*)

4. The rear of the Isotta's chassis, where the spring attaches, is typical of cars of this period (*Simon Moore*)

engine in the 1990s. It is probable but not certain that this was the FENC registered 1041 in Victoria and pictured in the photograph of the Melbourne dealership when new.

Cresswell died about five years ago and the car was sold by tender in October, 2008, to Warwick Anderson and Peter Latreille in partnership. The new owners were keen to do a full restoration and Beverly Greig and Ian Morrison agreed to allow them extensive access to the other car in Melbourne. In addition, all agreed that Ken Styles should oversee a lot of the work using his knowledge of the other car to the maximum.

I was fortunate to be able to see this car during its restoration both in July, 2011, with Peter Latreille and early in 2012 with Ken Styles and Bob King. The work was almost finished and indeed the car was running by March, 2012, with a body copied from the Duckett car. Since then, the owners have covered almost 1000 kilometres on the roads of Victoria. It is almost incredible that these two cars have never lived outside the state of Victoria, and one is in only its second family ownership.

Peter sent me a description of the first major run for the car in early May, 2012, as follows: 'James Earl and I did several test circuits of 36km on Friday, mostly through icy wind and rain, with a creamy stout at the Bunyip Hotel in Cavendish just before dark to revive us. Ian Morrison and Noel Cunningham arrived that night ready for the Saturday adventure, both cars driving 125km each way from Dunkeld to Great Western. Warwick Anderson and Julia could not come until Saturday night when we had the inaugural Giuseppe Stefanini memorial dinner, finishing at 2:39am on Sunday. So James and I drove the adventure on Saturday through the Grampians Range. This was a true lesson in how to manage low developed horsepower in a voiturette racing car, balanced with closely spaced but high ratio gears – a struggle on steep and long inclines in the mountain range, but great fun when 'wound up'. We averaged 50km/h for an hour, and were timed by a lady following at close to 75km/h before Halls Gap, where a warm coffee was in order. In all, 250km on Saturday. Sunday was a Dieppe triangle from Cavendish to Coleraine, to Hamilton, back to James and Jo's farm out of Cavendish: 110km. It was quite a revelation for the annual Bugatti Club Australia rally, particularly after all of the misconceptions about these cars over so many years. Bravo, Stefanini! The gearchange is the most delightful of any car I've ever driven, and the lightness of steering and brakes is remarkable. Roadholding, even on bumpy roads, is beyond reproach. The driving technique is, as for all small cars, one that uses the gears a lot, and that is how we tackle anything that is a noticeably upwards incline; anything that is a noticeable decline is flat-out for the next. Please find me another 1908 motor car of 1300cc that will perform in such sprightly fashion as the two little Isottas, both of which we all got to sample over the weekend. Total distance was exactly 460km and extraordinarily trouble-free. Full marks to Ken Styles and all that he did, especially to the engine and transmission.'

How many were made?
Firstly, it is clear that the FE engine was sufficiently different from that in the FENC

Identifying the survivors by their marks

	Duckett	Latreille/Anderson	Caruso/Mittler	Braen/Tavoletti
Dash plate	6006	-	-	-
Gearbox gate casting	6006	6017	6023	?
Engine	4 Zucco	3 Frangini	32 Ghezzi	Scripps-Booth 3005
Gearbox	1 Reschigna	6 Reschigna	10 Reschigna	?

1. The two Australian cars together after the restoration of the second car. From the left: Warwick Anderson, Peter Latreille (with their car) and Ian Morrison with the Duckett car (*Peter Latreille*)

2. Unknown location in the USA before World War One. This is the picture made famous by its inclusion in Ralph Stein's book *The Great Cars*. The Isotta in question appears to be an FE (*Tito Anselmi Archive*)

3. The Duckett car in the foreground and the Anderson/Latreille Isotta behind (*Rob Imhoff*)

that the engines were numbered separately, especially since one existing car has engine number 3. The same applies to the gearbox, which was only three-speed on the FE and where another survivor has number 1. As far as the complete car numbering is concerned, the survivors are all in a series starting 60xx, but whether the FEs were numbered 6001 to 6003 and then the FENCs came after that is unknown.

How many Type FENCs were there? The only clues we have are to look at the numbers on the various main components, since all but one car has lost its original Isotta Fraschini plate. Luckily, the number is also stamped into the gear lever gate.

The machinist or assembly engineer obviously stamped his name at the same time as the number. So we can conclude that at least 23 FENCs were probably made (depending on whether or not the FEs were in the same 60xx series) and probably 32, judging from the crankcase number, although the total number could have been more since the intention originally seems to have been to make 100.

The author would like to thank (in alphabetical order) Gordon Barrett, Dean Butler, Sébastien Faurès, Bob King, Peter Latreille, Miles Morris, Ian Morrison, Neil Murray, Doug Nye, Bob Rubin, Guillermo Sanchez, Norbert Steinhauser, Ken Styles, Jeffrey Vogel and George Wingard for their help in preparing this article.

Bibliography
- *Isotta Fraschini* by Tito Anselmi, Milani/Motor Books International, 1977
- *Racing Voiturettes* by Kent Karslake, Motor Racing Publications, 1950
- *The Great Savannah Races* by Julian Quattlebaum, self published, 1957
- *Bugatti by Borgeson* by Griffith Borgeson, Osprey, 1981
- *The Brescia Bugatti* by Bob King, Peleus Press, 2006
- *Ettore Bugatti, L'Artisan de Molsheim* by Norbert Steinhauser, Bugatti Book, 2008
- *A Record of Grand Prix and Voiturette Racing, Volume 1 (1900-1925)* by Paul Sheldon, St Leonards Press, 1987
- *The Golden Age of the American Racing Car* (second edition) by Griffith Borgeson, Society of Automotive Engineers, 1998

FE and FENC: the Isotta Fraschini voiturettes

Driving the FENC

Douglas Blain enjoys an experience few can boast of

I have lusted after one of these charming little 1402cc voiturettes ever since stumbling across the unforgettable illustration of one taken in a New York street before WW1 and reproduced across a double-page spread in Ralph Stein's *The Great Cars*. By comparison with rival designs of its era, the car is just so well proportioned and logical in its layout one somehow assumes it will perform and handle beautifully. And now I know it does.

Thanks to Melbourne estate agent Warwick Anderson, one of the co-owners, I was invited to try chassis number 6017 early one spring morning during my most recent visit Down Under. Like many of his fellow enthusiasts in that city of old-car fanatics, Warwick keeps his collection in converted artisan premises in Richmond, a gritty 19th century inner Melbourne suburb. Surrounded by full-sized exotics of various types and ages, the Isotta looked toy-like, delicate almost to a fault. Its newly completed restoration had been carried out very much in the spirit of its time, with nothing over-polished and that dull, lustrous sheen to the paintwork that results only from painstaking hand application.

Rather than resorting to the starting handle, Warwick prefers a procedure that he feels is kinder to the car as well as to his wrists (*pace* Gerry Michelmore, p48). Jacking up one back wheel, he lowers it onto a roller device like those used for starting racing motorcycles, powered by an adapted angle grinder. A lead runs from this under the car to the driver's side, where he can operate a treadle switch whilst manipulating the throttle and advance/retard levers.

The tiny engine came to life almost at once, with an eager bark from the exhaust and noticeably little extraneous mechanical noise. A minute or two sufficed to warm it through, whereupon Warwick climbed in behind the wheel and motioned to me to take the passenger seat. Without fuss or drama, we moved off into the teeth of a sudden rainstorm...

The narrow, slippery back streets in that part of Melbourne are studded these days with tiresome sleeping policemen, or mini-roundabouts. It was by now rush hour, so these kept my chauffeur busy with stops and starts in second gear while the rain found its way down our necks and up our trousers. Once out on Richmond Boulevard, however, a leafy green avenue of smooth, winding tarmac running alongside the River Yarra, he stopped the car and suggested I take over.

Comfortably seated (the driving position is perfect, with all controls nicely weighted and sensibly positioned), I took a moment to run through the four-speed gearbox with the multi-plate clutch disengaged just to reassure myself that the clearly defined gate was free of traps for the unwary. Its layout is in fact perfectly conventional and the ratios high and closely spaced, so one doesn't hesitate to use the box to row the little car along, keeping it up on the cam and allowing the heavy flywheel to build up and hold momentum.

From this you will gather that the Isotta's performance is hardly scintillating. With the car's weight at 740kg and, at most, 15 or 16bhp on tap at maybe 2000rpm, one would be foolish to expect too much. But, like an early 2CV Citroën, for example, of 40 years later, it seems to accumulate speed over time, with the result that one is constantly on the lookout for ways to avoid getting stuck behind traffic one considers to be slowing down unnecessarily.

The little car rides well, steers beautifully and stops without drama, so that driving it is a joy despite the lack of acceleration. Already the joint owners have done several long runs without trouble, although an entry in the famous Rob Roy hill climb had to be aborted. These were, after all, specialised racing machines designed for long, flat circuits with infrequent bends, only later adapted for use on the open road.

The First Motor Museum
a centenary

With the aid of rare photographs, **Michael Ware** turns the clock back to 1912 and describes the formation of the very first motor museum and its subsequent history

I have always been an admirer of Edmund Dangerfield, the man who had the foresight to collect cars for this country's, and possibly the world's, first motor museum. It opened in 1912 in Oxford Street, London, and later was transferred to the Crystal Palace. When the First World War came the cars were put in storage, and they had been dispersed by 1926. The museum may not have had a happy history, but it was a milestone at a time when major museums did not seem to be interested in the burgeoning of the motor car.

Edmund Dangerfield (1864-1938) was the proprietor and Editor of *The Motor*. He also published *Cycling* (he was a keen cyclist and collector of bicycles), *Motor Cycling*, *Commercial Motor* and *Motor Boat*. In 1910 he published in *The Motor* his ideas for a Museum of Motoring for Great Britain. No doubt he was inspired by the temporary display of some pioneering cars at the 1909 Imperial International Exhibition at the White City. Dangerfield brought together a group of trustees to help him with the choice of vehicles and the running of the museum. They were chaired by HRH the Duke of Teck, while much of the work of collecting the cars and the administration was undertaken by the Secretary, E S Shrapnell-Smith. Whilst the term trustees is mentioned many times, I think this was just a group name for the committee; I don't think they had 'a responsibility in trust', but I do feel they acted diligently when it came to collecting for the museum, and dealing with its later dispersal.

Some 40 pre-1904 vehicles were brought together for display on the ground and first floors of Waring and Gillow's premises at 175 to 179 Oxford Street, London. Temple Press, publishers of *The Motor*, produced a well illustrated and descriptive catalogue and the introduction to that publication is, I believe, worthy of repetition here:

'The nucleus of the collection, which it is hoped may be put on a permanent basis in the near future, is now provided in the exhibits which appear in this preliminary catalogue details. The whole of the expense and responsibility for various guarantees in connection with the Motor Museum are borne by *The Motor*. It has been a matter of common knowledge, for some years, that cars and cycles of great historical interest were being destroyed by their owners, or sold for breaking purposes. The fruitful period of experiment in modern road locomotion may be fixed between the years 1890 and 1899, yet, although less than 20 years have elapsed since investigations and inventions became prolific no responsibility exists even now of preserving many vehicles which attracted world wide attention by their comparatively recent performances.

'It might reasonably have been concluded that one or other of the Government museums would have been available for the purposes in view, but the fact remains that no action was taken, and that considerable difficulty has been experienced in arranging for a very small amount of accommodation – sufficient for only three cars at South Kensington [the Science Museum]. Again when the matter of definite action was brought up by Sir J H A Macdonald KCB at the Automobile Club, some years ago, nobody attended the meeting to hear the proposals and nothing was done. This state of inaction threatened to continue indefinitely.' The reason given for only three cars being taken for display in the Science Museum was a lack of space.

I cannot find any figures to show how many visitors the museum attracted. The public admission was 1/-, equivalent today to £2.85, but more importantly it represented a 17th of the average wage for an agricultural worker. The price must have kept a lot of the general public away. According to *The Motor*, commenting on the chosen price, it was said to have been set to prevent 'the use of the premises as a refuge in bad weather for uninterested persons'.

Having opened on 31st May, the exhibition closed on 31st July. The vehicles were then put away, but I can find no record of where they were stored. The trustees wanted to re-open at another central London site but were unable to find one, so the vehicles remained in store throughout 1913.

Many cars were displayed much as received. This is the first Argyll car made in 1899. It is now in the Glasgow Riverside Museum (Motoring Picture Library, Beaulieu)

1. John Henry Knight of Farnham on his Knight car, now in four-wheel form, in 1896. Today the Knight is owned by the Science Museum and is on loan to the National Motor Museum at Beaulieu (*Motoring Picture Library, Beaulieu*)

2. The exterior of the Waring and Gillows building in Oxford Street, as signwritten for the Motor Museum

3. The Bremer, which, according to the catalogue, 'ran on the road in December, 1894, but the body was not completed until January, 1895' (*Vestry House Museum, London Borough of Waltham Forest*)

4. The transfer from store to Crystal Palace was all done by horse-drawn transport. In the foreground are the 1897 Arnold Benz and the 1899 Renaux tricycle. The former is in the Royal Scottish Museum, Edinburgh, and the latter in the Science Museum store at Wroughton (*Motoring Picture Library, Beaulieu*)

On 12th March, 1914, the museum re-opened at the Crystal Palace, a place of much more popular entertainment than Oxford Street. If contemporary reports can be believed, the building had seen better days. The vehicles were all transported from the stores to Crystal Palace on horse-drawn transport provided by Robert Park and Co. The museum was again properly laid out, with captions for each of the exhibits. It is quite clear that the trustees were on the look-out for more exhibits to display, though I cannot find any recorded increase in the collection in that year.

Then came the outbreak of war. The Crystal Palace was requisitioned by the Admiralty and the museum had to be packed up again and put into store. As we shall see, all the exhibits seem to have survived the war years, though there was some deterioration. Michael Sedgwick, in an article in *Veteran & Vintage* magazine in November, 1959, wrote, 'One or two of the exhibits languished on a disused lot near Charing Cross Station until they were reduced to produce.' I would love to know where he found this reference.

Post-war years were not the best time to re-start a motor museum and the trustees realised there was now little hope of being able to keep the collection together. An approach to the Science Museum in 1919 saw a little relaxation of their previous view, saying they could take seven cars but even then could not guarantee there would be space to show them together. Approaches to other museums brought very little response. Hull offered to take three, Glasgow two and the Royal Scottish Museum in Edinburgh four. By 1922 the trustees were getting desperate. A new regime at the Science Museum agreed to take the rest of the collection and dispose of them as appropriate. That great acquisitive curator, Tom Shepherd, at the newly opened Museum of Commerce and Transport in Hull agreed to take another six. A lot of research has been done to find out where the cars are now, but the actual disposal routes of some are not too clear. It would appear that the Science Museum, who had said they would not keep unrestored cars, did scrap three, though parts from each were retained.

In 1987 – the 75th Anniversary of the Motor Museum – the National Motor Museum at Beaulieu staged a small retrospective exhibition relating to that first museum, putting together six cars from the Oxford Street exhibition. Four were normally on show at Beaulieu, but the Science Museum kindly loaned their 1898 Darracq and the 1899 Renaux tricycle, both displayed very much in the same condition as 75 years earlier. It is thought neither had been on public display anywhere since 1914. On the opening day the exhibition was visited by Richard Dangerfield, grandson of Edmund, and John Thorpe, Assistant Editor of *Motor*.

The Motor Museum catalogue of 1912 lists 37 three- or four-wheeled vehicles, two motorcycles and a Clarke bicycle wheel and engine. Many of the illustrations have been heavily retouched to improve the look of the vehicles, whilst some of the photographs taken at the time show them in their true form. The catalogue makes interesting reading, with between 200 and 250 words dedicated to each vehicle. It is mostly technical, which makes one pretty sure that the museum was intended for the existing or future motorist and not the general public. There are traces of humour, however. I have chosen to comment here on just a few of the

The First Motor Museum – a centenary

1. The 1894/95 Cannstatt Daimler in reasonably sound looking condition. It is now in the Science Museum store at Wroughton

2. The 1900 Wolseley, now with British Motor Heritage at Gaydon

3. The 1900 Albion, the second Albion made. Used by the first owner for 11 years, it is now in the Royal Scottish Museum, Edinburgh

4. The 1900 cyclecar, believed to have been built by Henry Sturmey and now in the Streetlife Museum, Hull

5. The 1901 Delahaye which was still on the Science Museum's books in 1935. Declined at that time by Hull Museum, it was scrapped in 1936

6. A steam tricycle built before 1900. 'The history of this vehicle is a complete blank…' It had disappeared from the museum collection before the Science Museum took over the disposals. The same fate appears to have overtaken another early steam tricycle from the museum, patented in 1881 by Sir Thomas Parkyns and made by Arthur H Bateman of East Greenwich. Does either survive?

7. The 1897 Pennington tricycle. This is the only known survivor of the production of Penningtons. It is now owned by the Nash family and exhibited at the National Motor Museum at Beaulieu

8. The 1897 Lanchester, the second one built and the oldest surviving, now in the Science Museum collection

9. The 1900 Cleveland Electric, the last car listed in the catalogue. It is now in the Streetlife Museum, Hull

10. The 1897 Bersey electric motor cab, showing the easily detachable tray of batteries. It is now in the Science Museum collection

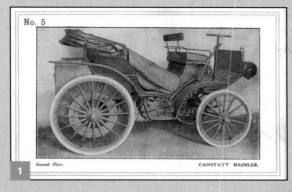

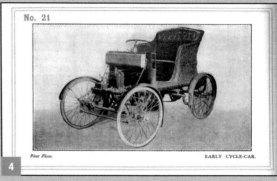

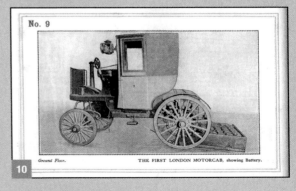

1. Described in the catalogue as 'an early Turgan car, probable date of construction 1899'. In hindsight a very rare and mechanically interesting car, it's a pity it was later scrapped as no-one wanted it (*Motoring Picture Library, Beaulieu*)

2. The 'Benz' as found in a wood in Berkshire. The engine had been removed and was powering an electric light plant in a nearby house

3. The 1894 Cannstatt Daimler en route to the Crystal Palace, Robert Park and Co in charge (*Motoring Picture Library, Beaulieu*)

vehicles on display; quotes come from the catalogue, as do all dates mentioned.

1892-94 Bremer: Presented by the Bremer Engineering Co. 'This car was built by Fred Bremer, of Walthamstow. It ran on the road in December, 1894, but the body was not completed until January, 1895.' It was returned to its builder c1922. On his death in 1941 it was presented to the Vestry House Museum in Walthamstow, where it is now on display. The Bremer was displayed in a ground-floor window of the Motor Museum, along with the Pennington.

1894-95 Cannstatt Daimler: An open landaulet complete with weatherproof sheet to protect the passengers' lower bodies, it was lent by C E Grey of the Golf Links Hotel, Hunstanton. It 'carries the two-cylinder vertical engine at the back, where it is enclosed in a sort of 'meat safe'....a good example of the original hot-tube ignition, although the platinum tubes have disappeared – they were worth about 30s each once; more lately.' This exhibit passed to the Science Museum and is in store at Wroughton.

1900 Wolseley: Lent by the Wolseley Tool and Motor Co Ltd, this was one of three early Wolseleys loaned to the Museum and was that company's first four-wheeler. It was very successful in the 1000 Mile Trial, and in 1950 was driven by St John Nixon from John O'Groats to Lands End. It 'is in truth the parent of all the bonneted Wolseleys – a bonnet which many visitors will remember was largely full of 'emptiness', and consisted of neatly ranged stacks of gilled tubes.' These three cars were later returned to Wolseley and are now part of the British Motor Heritage Collection at Gaydon.

1897 Bersey Electric motor cab: Presented by Mr W C Bersey, this was the first motor cab in London. A fleet of 25 was running by end of 1897 and a further 50 by the next year. All had gone by 1899. The illustration in the catalogue is interesting as it shows a set of batteries about to be fitted, and the car was displayed with battery box underneath. This is now in the Science Museum collection in store at Wroughton, having previously been on display at the National Motor Museum at Beaulieu for many years.

1898 English-built Bollée: Lent by *The Motor*, it 'was built in the Old Conduit Yard, Coventry, by a company run by Harry J Lawson. M Bollée got £10,000 in cash for his patents – those were the halcyon days for investors – and he could not believe his good fortune at first, and was not easy in his mind till he had the cash in his pocket.' This vehicle is now part of the Streetlife Museum in Hull.

1899 Turgan, made in France. Lent by *The Motor*, 'this vehicle was for some years used by the Rev John Swimnnerton of Tolleshunt Major, Witham in Essex, a blind clergyman of the Church of England, who employed a youth to drive him about his parish.' This is one of three cars broken up by the Science Museum in November, 1931, when they could not find any museum to take them on. Certain parts were retained in the collection.

1895 Knight: Lent by John Henry Knight of Farnham. 'Completed and ran in July, 1895. Originally a three-wheeled vehicle. Constructed at Barfield, Farnham, Surrey. Altered to a four-wheeler in April, 1896, and ran at the Crystal Palace Exhibition, May, 1896. The one-cylinder engine was designed and made by Mr Knight. Hot-tube ignition effected the firing of the charge originally, but electric ignition (by battery, coil and plug) was substituted afterwards.' Returned to J H Knight, who in turn presented it to the Science Museum in 1957. It has for many years been on loan to the National Motor Museum, Beaulieu.

1899 Star dogcart: Presented by the late Eric J H Shepherd. 'A steering lever of less magnitude than some present day throttle-levers was provided, and a pointer gave one some indication as to the probable direction of the car. If one wanted to reverse, one had to go on wanting – or else get out and push.' In 1924 this car was sent to Malcolm Campbell's London showroom for display. It is not known if this was a loan or gift. It was never returned to the Science Museum, and is now believed to be the Star on exhibition at Thinktank, Millennium Point, Birmingham.

'c1899 Benz': Presented by *The Motor* and described in the catalogue as a 'disreputable exhibit', this car could not at that time be positively identified. It had been found by 'an inspector in a copse in Berkshire, forming an apparently comfortable resting place for a healthy community of poultry.' It is not recorded as having reached the Science Museum's disposal team. It disappeared and only emerged in 1995 at a Christie's sale, described as a 'Veteran-type vehicle'. It was bought by Tony Taylor, deeply researched by Malcolm Jeal and found to be an Endurance, made in Coventry.

1899 de Dion-Bouton Quadricycle: Presented by the Earl of Carnarvon. 'The engine is in a position behind the back axle. Such a design was known on occasion to result, when starting, in the complete somersaulting of the machine. Provision is made for assisting the engine by means of pedal driving.' Despatched to Hull Museums in 1924, this Quad can be

The First Motor Museum – a centenary

4. The 1899 Star, now on show at Thinktank at Millennium Point, Birmingham. One does occasionally wish curatorial sense would prevail over some ideas coming from museum designers

5. The 1895 Cannstatt Daimler, now in store at the Science Museum, Wroughton

6. The 1899 de Dion-Bouton Quadricycle. Taken to Hull Museum in 1925, and now displayed as part of a reconstruction of a blacksmith/cycle repairer's workshop in the Streetlife Museum

7. The 1898 Darracq, lent originally by the makers. Note the unusual colour scheme; the paintwork is almost certainly the original. According to the politically-incorrect catalogue it was 'air-cooled, with the cooling vanes carefully arranged so that the front ones shielded those behind from the air draught! Perhaps the designer had a holiday in Ireland before designing the engine!' It is now in store at the Science Museum, Wroughton

viewed today in a reconstruction of a blacksmith and cycle repairer's workshop in the Streetlife Museum, Hull.

1900 cyclecar: Presented by Henry Sturmey, the first Editor of *The Autocar*, it is thought that Sturmey built it himself. 'This machine is believed to be the first cyclecar with the special object of producing a car that could be sold for £100. This car did 15mph and its designers were satisfied. It would take a one-in-eight grade on its 'first', but then one had to stop to cool the engine, which got so hot that it went on firing when the spark was switched off.' Despatched to Hull Museum in 1925, it is now on show in Streetlife.

Prior-to-1900 steam tricycle: Presented by Sir Charles Friswell. 'The history of this vehicle is a complete blank. It came into the possession of Sir Charles Friswell 10 years ago, but he knows nothing of its early life or manufacture.' The only record we have of its fate after the Motor Museum closed suggests that it did not reach the Science Museum's disposal panel. The same fate seems to have befallen another steam tricycle, the one built by Arthur H Bateman of East Greenwich for Sir Thomas Parkyns in 1881 using his patents. Does either of these very interesting looking machines survive today?

1897 Pennington tricycle: Lent by C A Smith, a famous cyclist, of the White Lion Hotel, Cobham. The history of Pennington and his antics are well known. The catalogue claimed 'It was very fast – for those days. It would do about 40mph, and was undoubtedly the fastest thing on wheels at the time – while it ran.' It almost sounds as if Pennington himself had something to do with the writing of this entry. When the museum closed the Pennington was returned to Mr Smith and I have seen a picture of it in the 1920s or '30s at his hotel. It later passed to R G J Nash, who had a collection of early cars and bicycles at Brooklands. It is now on loan from the Nash family to the National Motor Museum at Beaulieu. It was originally displayed in the window of the Motor Museum with the Bremer.

1896 Lanchester: Presented by the Lanchester Motor Co. 'The original Lanchester car. A six-seater phaeton, with 8hp engine. Constructed at Birmingham, and all parts made in England.' A very important car, I have often wondered why it did not take part in the original Emancipation Run. It was returned to its makers in 1924. Sadly it was destroyed in the Coventry Blitz. The second Lanchester, built in 1897, was also displayed at the museum, and that was later accepted by the Science Museum.

Not included in the catalogue are four other vehicles, which means they either arrived at the museum after the catalogue went to press or, more likely, were added to the collection in 1914 at the Crystal Palace. Very sketchy information is known, but they were a US Long Distance, which was on the disposal list in 1922 but had no takers and was scrapped in 1931; an 1899 Lawson Motorcar and governess car, given to the Science Museum by its owner in 1922 but disposed of in 1933 (is its fate known?); a Fouillaron Brown tonneau passed through the disposal board in 1926 to City and Guilds College 'for a mascot' (could this be the James and Browne now campaigned every year on the Brighton Run by Imperial College?); and a large 1903 Serpollet, now in the Science Museum collection.

That first attempt at creating a motor museum for Britain was a brave experiment. Though in the end it was not a very happy saga, I am convinced that many of the cars exhibited there would not have survived if it had not happened. As it turns out, the first National Motor Museum for this country did not open until 1972 – 60 years after Edmund Dangerfield first put cars on display in Oxford Street.

I must acknowledge the help I have received from consulting David Jeremiah's article 'The formation and legacy of Britain's first motor museum' (Journal of the History of Collections, 10/1) and Michael Sedgwick's article on the museum in Veteran & Vintage, November, 1959. Michael Worthington-Williams, Peter Mann and Malcolm Jeal have done a great deal of work in recent years in tracking down the fate of many of the cars from the museum. I have drawn heavily on their research.

The Trustees
HRH the Duke of Teck (Chairman), Sir David L Salomons, Hon Arthur Stanley, Sir Boverton Redwood, Sir J H A Macdonald, Col H C L Holden, Edmund Dangerfield, Claude Johnson, W Joynson-Hicks, Julian W Orde,
E S Shrapnell-Smith (Secretary)

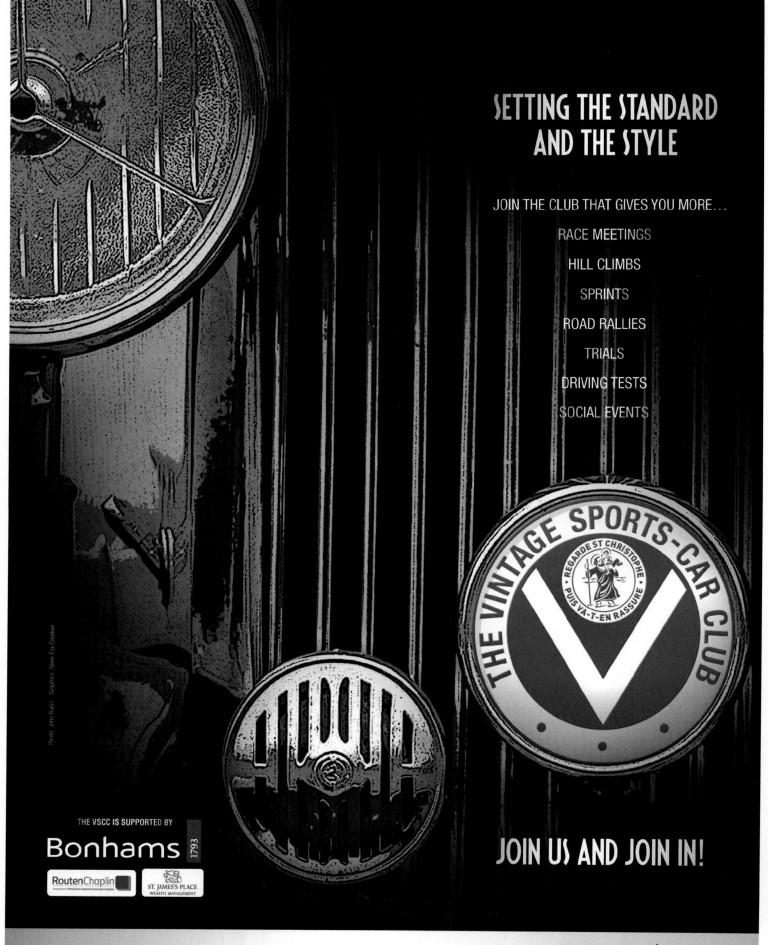

Marseille maverick

standstill.'

Rougier was on a roll, and had his Voisin crated and sent by train to Rome for September's meeting at Brescia, the first in Italy. As in Reims, high society set the tone. Among the dignitaries to whom he was introduced were Victor Emmanuel III, Princess Letizia Bonaparte, Puccini and the assiduously self-promoting d'Annunzio, who posed happily by various craft for local Kodak enthusiasts. Uneven ground made landings hazardous at Brescia. The crash-prone Blériot flew into a tree, Mario Calderara smashed his Wright's rudder and Allessandro Anzani broke the prop of his Italian-built Voisin, which was promptly cannibalised by Cagno for his similar Itala-engined machine. On the opening day, Rougier, experimenting with a new Rebus engine, made a flight lasting more than 12 minutes and beat Glenn Curtiss by reaching an altitude of some 100m the next day. Skipping around in delight when his victory was announced, Rougier enthused: "Curtiss is a true gentleman. He could have gone higher than me, but he promised to leave me the first prize and kept his word. This is true American chivalry!"

Among the spectators was Franz Kafka, holidaying with Max Brod in nearby Riva. He described Rougier as 'sitting at the controls like a gentleman at his desk', as if to equate aerial endeavours with the soaring flights of his own imagination. Given the great heights Rougier attained: 'He appeared so high in the air that one thought his course would soon have to be determined by the stars that were about to show themselves in the darkening sky,' he wrote. 'We couldn't stop turning around; Rougier was still climbing as we drove away into the *campagna*.'

Thenceforth, Rougier made high altitudes his metier. The aeronauts' caravanserai decamped from Brescia to Berlin for the International Flying Week at Johannisthal, where Rougier let it be known that he intended to better the 155-metre record Latham had set at Reims – perhaps even Orville Wright's unofficial 172 metre ascent of 10 days before. Although he took the distance award, with 31 laps covering some 77 kilometres in 97 minutes, Rougier failed on both counts. He did, however, win £3200 in prize money, as well as conducting the world's first aerial interview with the eminent aeronautical journalist Georges Prade as his passenger.

Contemporary reports illuminate Rougier's competitive character. Having arranged to collect a new ENV V-8 from Berlin's main railway station, he arrived to find only the crated unit destined for the Voisin of his former Mercedes-driving adversary Pierre de Caters. Completely disregarding such trifles as ownership and delivery instructions, he whisked the engine away and installed it in his own aeroplane before the Belgian baron could object. He was stereotypically Mediterranean, too: German press reports describe Rougier as 'a small man with a remarkable nose, pacing around his hangar in a state of extreme agitation, constantly waving his arms and hands, grabbing his own body everywhere, ordering his crew this way and that,' while his imperturbable wife saw to the business side of things. But it was this devil-may-care recklessness that made Rougier such a favourite with the crowds. In contrast to Farman, who prudently hugged the ground, the mercurial Marseillais always thrilled by climbing as high as his machine would allow. In Frankfurt the following week, he flew above the dirigible *Parsifal III* in a bold manoeuvre that sent the gleeful crowd wild with hat-throwing excitement – especially as he then cut his engine at an altitude of some 75 metres and glided down in what *The Aero* described as 'an exceedingly graceful curve, while the people below, thinking the aviator had lost control over his motor, stood aghast in apprehension of a terrible accident.' Rougier knew how to put on a show.

From Germany, the Voisin was shipped to Doncaster in late October and then on to Blackpool for the first British air meets, where he repeated his Frankfurt 'no engine' stunt: 'The magnificent spectacle of an uninterrupted hair-raising glide from 200 feet, the machine a patchwork of light and shade against a white cloud, was cheered to the echo,' cried *The Aero*. For a man with expensive tastes, this relentless schedule was

accumulating a very satisfactory prize pot. Blackpool netted him another £530, and in Antwerp the following week he won the 25,000-franc Grand Prix for the longest flight, the 5000-franc altitude prize (270 metres), a 2500-franc bonus for beating the world altitude record in the process, plus five per cent of the gross takings for the whole event – more than enough to buy himself a new Voisin. Which he did.

At the end of the season, the Aéro Club de France granted Rougier pilot's licence number 11. Soon after the New Year, he accompanied his American/Belgian de Dietrich team mate and former Land Speed Record holder Arthur Duray, Jean Gobron (of Gobron-Brillié), Mortimer Singer (of the sewing machines), the Voisin brothers and the world's first aviatrix, Elise Deroche, to Cairo for the Héliopolis Aviation Week, to compete for nearly 300,000 francs of prizes. Although the event was dogged by high winds, some 40,000 spectators (including the Khedive's enormous harem in a separate marquee) saw Rougier win the Grand Prix d'Égypte for the greatest total distance flown – some 153 kilometres. His flight over the Great Pyramid of Gizeh was subsequently immortalised in a popular pochoir print by Marguerite Montaut.

In March, 1910, Rougier accomplished the feat that would secure his place in aviation history. After a disastrous attempt to host a meeting the previous year (in which no one flew), the Principality of Monaco offered a healthy purse for the first to fly out over the Mediterranean from the old port and ascend La Tête du Chien, the 550-metre outcrop that overlooks Monte Carlo. It was a challenge Rougier couldn't resist, and thence his équipage duly sailed from Cairo.

The landing strip was on Quai Antoine 1er, which stretches some 400 metres seawards from the Rascasse corner of today's F1 circuit. Even with sandbags piled up at one end and a crude braking system improvised for the Voisin, take-off and landing required luck as well as precision. With a 20-metre wall to be cleared at the end of the quay and a tight turn executed immediately afterwards before climbing steeply to round Cap Martin, more experienced aeronauts such as Farman and Wright – even the indomitable Blériot – refused even to try. The quayside was packed, Monegasque rubberneckers even festooning the florid cast iron lamp standards lining the narrow promontory, when the Voisin's stubby exhaust stacks barked into life. Rougier, immaculate as ever in a long, tailored blazer, high celluloid collar and flamboyantly broad flat cap, his patent leather shoes glinting in the afternoon sun, climbed the stepladder into the cockpit and nodded his readiness to Gabriel Voisin for the crew to let go of the trembling machine. In less than 100 metres he was aloft, the blare of the big V-8 under load drowned by the roar of the crowd as it speared its wobbly way across the bay to Cap Martin.

The world's first flight over sea was clearly no picnic, however: "I left in a calm wind," Rougier told journalists later. "Crossing the Cap d'Ail, I was struck by gusts so violent that my machine stopped in mid-air before being hurled hither and tither like a child's rubber ball. Expecting to be dashed into the sea, I was suddenly thrown 200 metres into the air, completely out of control. I tried in vain to turn to escape the wind. After several attempts I finally headed downwind, travelling at over 100km/h towards the Bay of Menton where I hoped to ditch in the sea, but, by the grace of God, I found calmer winds and managed to land safely. In 15 minutes I experienced total fear and a profound instinct for self-preservation. On landing on terra firma I was soaked in sweat despite the icy wind, arms bruised, breathless, happy and relieved."

After Monaco, Rougier fitted his new Voisin with one of the state-of-the-art Gnôme Oméga rotaries that had so convincingly proved their superiority with Louis Paulhan and Henry Farman the previous summer. Thus equipped, he spent April with varying degrees of success at the meetings in Florence (where a violent sideways gust on his final approach flipped the machine over and wrenched off the right wing), Padua and Nice, where rudder failure forced him to ditch into the sea and, entangled in bracing wires, he narrowly escaped drowning. Despite assurances at his hospital bed that evening from the suave Hubert Latham, who had done the same thing the previous year in the less clement waters of the Channel, that such adventures were "all part of the game," the incident shook Rougier. He had two more commitments to honour: in May he returned to Berlin, where he flew with two passengers before heading off to Hungary, where the Budapest meeting offered the biggest prize pot of the season. The highest casualty rate, too: the Voisin of Juan Bielovucic crashed heavily, André Frey's Sommer was blown into a crowd of spectators, causing many injuries, and Latham's Antoinette and Louis Wagner's Hanriot were reduced to matchwood; the local pilot Zosely lost his life.

Enough was enough. As the fourth highest earner of prize money from aviation and feeling lucky to be alive, Rougier decided the odds were no longer in his favour. An aristocrat of the air, he was already famous throughout Europe. Gratifyingly for a gambler, he now numbered a royal flush of crowned heads (and quite a few knaves) among his acquaintance, and had the business connections to maintain his Champagne lifestyle. He therefore announced his intention to give up flying to concentrate on his motor businesses and other interests.

Unsurprisingly for one so entrepreneurial, these included aeroplane manufacture. Sensing a potentially lucrative new market among wealthy sportsmen, Turcat and Méry had provided a lightened version of their 18hp four-cylinder automobile engine to the Vendôme brothers and their young engineer Alexandre Odier to power their pretty and exceptionally light pusher biplane with eyebrow wings. The Turcat Méry & Rougier company was formed in Levallois-Perret to develop this promising design, with a slender new empennage and a tractor propeller driven by the punchy big ENV eight-cylinder. Ingeniously, the tiny landing wheels were set into bifurcated steamed willow arches elastic enough to amortise landing shocks. The TMR was exhibited at the 1910 Salon de l'Aéronautique, and flew several times. Rougier told *The Aero* in September, 1910, that this "highly original machine" was "extremely fast", but it progressed no further. No more successful was his other ambitious project of 1910, a Pérignon-designed compact 8.5-litre four-cylinder 60hp aero engine, of which nothing more was heard.

Today, Rougier is best known for an achievement he dismissed as unimportant at the time: winning the first Monte Carlo Rally. Local hoteliers had benefited greatly from the publicity and glamour that the principality

Marseille maverick

1. For the first 250 miles of the 1923 French Grand Prix, Rougier averaged 81mph in his Voisin *Laboratoire*

2. Rougier drove his Turcat-Méry to second place in the 1921 Circuit de Corse

3. Outside the Voisin factory at Issy, Rougier poses in the four-litre he later drove to victory in the Touring car class at the Strasbourg Grand Prix in 1922

4. Storming les Alpilles in May 1922, Rougier took first place in both Racing and Touring car categories in the first Voisin Laboratoire, a four-litre with a lightweight aviation-inspired body

5. A worried-looking Rougier with his mechanic Lalauric in the monocoque Voisin C6 *Laboratoire* before the 1923 Grand Prix de l'ACF at Tours

had attracted by hosting the world's premier powerboat meetings and the aviation events which soon followed in the wake of Rougier's first flights, and a trans-European motor rally was seen as yet another way of trumping Nice and the other Côte d'Azur resorts. Funded by the Société des Bains de Mer and organised by the Club Sport Automobile et Vélocipédique de Monaco with the support of Rougier's new chum Prince Albert, the event required entrants to bear prominent Rallye Automobile Monaco plates from their various cities of origin: two from Geneva, nine from Paris, one from Boulogne, two from Vienna, four from Brussels and two from Berlin. The competitors were awarded one point per kilometre per hour average up to 25km/h, one for every 100 kilometres driven, two for each passenger (including mechanic) and up to 10 points for the comfort in which they travelled. The cars were then rated out of 10 for elegance, for how much the judges liked the coachwork, and for the condition of the chassis on arrival. Rougier started from Paris in a 25hp Turcat-Méry saloon that averaged 23km/h, and despite (or because of) this Byzantine scoring system he pocketed 10,000 gold francs as the victor.

Until the unfortunate Gräf & Stift incident at Sarajevo, Rougier devoted himself to expanding his dealership empire, using his celebrity and society networks to promote the increasingly glamorous big Turcats, following the arrival of the handsome MJ chassis. When the contract between Turcat-Méry and Lorraine-Dietrich was dissolved, he had taken over the Turcat agency in Paris, with glitzy showrooms in a prime spot on the Champs-Élysées in addition to his rue de Cormeille dealership in Levallois. He also entered into partnership with the engineer Robert Guyonnet to form a new company, Guyonnet & Rougier Automobiles, Aéroplanes, which registered the design of a neat folding head that disappeared completely from view beneath a hinged panel. This they applied to the astonishing Turcat-Méry *Squale* (shark) for an English client. With huge headlamps incorporated into the leading edges of the front wings, it looked more like a startled tree frog than a fish; the wings themselves, however, were smoothly integrated into the scuttle of this smooth, barrel-sided and round-tailed torpedo.

Rougier had a good war, initially in the motor transport service, then training pilots on Caudrons at Pau and Chateauroux from 1915 onwards. He was awarded the Croix de Guerre in 1917, and the citation for his Légion d'Honneur in 1920 commends his energy and sang-froid. The aftermath proved less rosy for Turcat-Méry. The much enlarged factory and workforce were now geared to the production line manufacture of munitions and commercial vehicles rather than the bespoke production of quality cars, for which there was in any case dwindling demand. Raw material supply was erratic, as was credit. While developing the 15hp sidevalve three-litre it managed to launch in 1920, the firm was therefore happy to sell Rougier its stock of redundant prewar chassis, which he updated with Perrot-braked front axles and motley ex-military aero engines, registered as Rougiers, and offered to his wealthy clientele. One such was the so-called 'automotive Raphael', which the French authorities forcibly repatriated from Charles Morse's Seattle collection in 2009.

Rougier assembled this magnificent hot rod – surely the ultimate Oily Rag Vintage tourer – for Ferdinand d'Orléans, Duke of Montpensier and brother of the pretender to the French throne. A longstanding customer, he had in 1906 commissioned Rougier's business partner Pérignon to build the first of a series of Class One Lorraine-Dietrich racing powerboats for the Monaco meetings. Then in 1908, Rougier sold the young Bourbon adventurer a reinforced 24/30hp Lorraine chassis on which he oversaw the erection of a double phaeton body by Berton-Labourdette, equipped with two vast trunks, a field kitchen (complete with table and chairs), folding beds, a generator and a tent. The nearside rear wing bore two acetylene lamps and an auxiliary fuel tank, and a hefty winch was mounted on the running board. Entry was by a single door, the other being modified to carry guns and ammo, spades and a pick; three axes were mounted on the radiator. The 3.7-tonne ensemble must have been pretty sturdy, as it survived being dropped several metres during transhipment en route to Saigon, from where the tiger-hunting owner forged his way overland to Angkor Wat.

The Duke's famous aero-engined leviathan of 11 years later was also destined for Indochina, and he optimistically ordered it in November, 1919, for delivery the next month. For the not inconsiderable sum of 33,500

francs, Rougier mated a 1914 MJ series 35CV Turcat-Méry chassis (on which he had enjoyed some success at the Mont Ventoux and Limonest hill climbs) to a 1915 Type Aviation Militaire 6A Lorraine-Dietrich 9.5-litre ohc engine. A compact and relatively light unit with six cylinders cast in three pairs à la Mercedes on an aluminium crankcase, it generated 110hp at 1650rpm and a gratifying surfeit of torque. By midsummer, however, the royal client still had no car. To Rougier's manifest irritation, Million-Guiet, whom he had contracted to build the handsome torpedo body, were not only dragging their heels but also demanding 5000 francs more than the 11,700 originally agreed. He even tried placing the work with Rothschild & Fils, but even for such a blue-blooded client they quoted a six-month delay. The subsequent story of the car is well documented, but its genesis betrays the chaotic state of the French motor industry in the economic turmoil of the immediate post-war years.

In a vain bid to revive Turcat-Méry's fortunes, Rougier once again took the wheel. He finished second in the 1921 Grand Prix de Corse and won his class at Mont Ventoux and La Turbie in 1921 and 1922, when he knocked more than a minute off the preceding year's record. Admittedly, this was bettered by an even larger margin the following year, but that was by René Thomas in the formidable five-litre Delage. Even with the injection of most of Léon Turcat's personal capital, the Société Marseillaise de Crédit pulled the plug. Turcat and Méry remained for a time as mere employees, and the company teetered from one optimistic buyer to the next for the rest of the '20s.

Rougier therefore sought other strings to his motor-coping bow. Of all the marques suited to his high-rolling clientele, Voisin was the obvious choice. Rougier had no connections with Hispano, Bugatti or Delage, the Farman brothers had their retail arm in-house, Panhards were too pedestrian and Delahayes too dull. By contrast, he had known Gabriel Voisin for years and shared with him a taste for the good life and, of all the debutant post-war makes, Avions Voisin had made by far the biggest splash. The President himself drove one, and with a list of owners that read like an admixture of the *Almanach de Gotha* and *Variety* magazine, the enterprise was very much Rougier's *tasse de thé*. Etablissements Rougier was duly established in the rue de Cormeille as the capital's fifth Voisin concession.

His first works drive soon followed, in an evolution of the old four-litre warhorse whose pace its co-designer, Ernest Artault, had already so impressively demonstrated at Gaillon and other hill climb events in 1920. It was the first racer to emerge from the Voisin Laboratoire, the experimental research department housed in an annexe not far from the factory and personally overseen by Voisin and his protégé, the young André Lefèbvre.

1. Built to contest the Touring car class of the 1924 Lyon Grand Prix, the ungainly 130bhp Voisin C9 *Laboratoire* was built around an aluminium and wooden monocoque

2. Rougier before his ill-fated last race, beside André Lefèbvre in the 10hp *Petit Laboratoire* iteration of the same concept

3. Hermann Rützler's streamlined 4.4-litre Type VI Steyr saloon in 1924

4. The famous Turcat-Méry that Rougier assembled for the Duke of Montpensier. It was powered by a 9.5-litre Lorraine-Dietrich aero engine (*Gooding & Co*)

Built in 10 days, the car featured a tapering long-tailed hull only 90cm wide, made of braced laminated hoops and doped linen and allegedly weighing less than 20kg. With skimpy wings, a windscreen that lowered into the scuttle, ingeniously minimal weather protection in the Voisin manner and an engine tweaked to deliver some 100hp, it proved fleet enough to propel Rougier to victory at the Pic Montaigu and Limonest hill climbs in May, 1922.

In July, at the age of 46, came Rougier's finest hour as a driver: the ACF Grand Prix de Tourisme in Strasbourg. Covering 53 laps of the triangular road course over some 700 kilometres, the race was for four-seaters carrying 70kg of ballast for each notional passenger and no more than 26.4 gallons of fuel and lubricant. The minimum dry weight was 1400kg, and the minimum body width, 1.15m. Just as the Kaiserpreis regulations had occasioned concave-sided 'tulip' torpedo bodies in Germany a dozen years earlier, Voisin famously accommodated this stipulation by adding faired bulges on either side of the otherwise slender C3 Sport team cars.

Rougier stood beside the second of the four works cars at the 7am start. Beside No 1 was Arthur Duray, the former de Dietrich team mate with whom Rougier had flown the Pyramids – a hard-charging racer respected on both sides of the Atlantic, his achievements including the Vanderbilt Cup and the 1914 Indianapolis 500 (for Peugeot). The No 3 car was in the hands of Richard Gaudermann, formerly Albert Clément's riding mechanic during the fatal practice session at literally breakneck speed for the 1907 Dieppe GP, and who had been successfully campaigning the four-litre Voisin Type Gaillon in hill climbs the previous year.

Oberto Piccioni was at the helm of the fourth car; the previous year he had driven for Steyr, which Rougier was representing in France (or was about to).

The combination of generous porting, Elektron pistons and unusually high compression furnished the standard double sleeve valve four-litre with a reliable 120bhp, which proved ample for taking on the Peugeot and desmodromic Bignan opposition. Trouncing the former was particularly gratifying for Gabriel Voisin, because not only was the 174S effectively a mirror-image clone of his own four-litre, but Artault, the defector who had jumped ship to Peugeot with all the drawings, was driving one. Although Gaudermann put in the fastest lap and Piccioni led briefly until he burst a tyre, Rougier crossed the line at an average speed of 107km/h with three litres of his fuel allocation remaining. With first, second, third and fifth places, the Strasbourg wins put the marque on the map, and were publicised accordingly.

Rougier looked his age by the time he insinuated himself into one of the ultra-light monocoque C6 *Laboratoires* in the racing car class for the French Grand Prix at Tours in 1923. Conceived and constructed in six short months, and with no experience whatsoever of building pure racers, these extraordinary beetle-like projectiles, as narrow as their aerofoil profiles were low-slung, had one singular disadvantage for the two stocky senior team drivers: the combination of anno domini and foie gras made it difficult for Rougier and Duray to get in or, once in, to see much out. Their two mounts were therefore fitted with squared-off steering wheels (although at Monza they made do with the circular variety). Not that this facility did the two veterans much good, as it turned out. The svelte Lefèbvre brought No 10 into fifth place at an average of 63.2mph – not bad for a hopelessly outclassed and seriously underpowered car (estimates vary between 60 and 80 horses) against far more potent factory opposition in a 500-mile race that claimed 13 of the 18 starters. Trying vainly to match the pace of Friedrich's Type 32 Bugatti 'tank', Duray and Rougier both retired, although Rougier averaged 81mph for the first 250 miles. The C6s also ran at the Italian GP in Monza a month later, in September; 'also ran' being the operative phrase, as the cars of

Rougier and Eugenio Silvani (another of his Italian Steyr drivers) both expired on lap 28, and Lefèbvre retired the lap after. If the C6 must ultimately be counted a failure, it was a heroic one. Its subsequent status is deserved, if only as a testament to the indomitable self-confidence that drove Issy's tight-knit little Laboratoire team.

Reckless as the experiment may have been, Gabriel Voisin was no fool; despite his skirmishes with the ACF bureaucrats, he decided to revert to the Touring car class for the following year's event at Lyon. While the new cars were being built, Rougier chalked up the first victory for the new four-litre C5 – the production version of the Strasbourg chassis – at the Estérel hill climb, near Cannes.

Rougier's propensity for wheeler-dealing proved his undoing at Lyon. The race was in two parts: an eight-hour endurance test run mostly at night, which provided the handicapping for the following day's speed trial, with strict fuel consumption limits. There were three touring car classes: A (more than 1400kg), B (up to 1000kg) and C (less than 400kg). The five-car Voisin team spanned all three classes: three dark blue C9 Grands Laboratoires in group A for Rougier, Gaudermann and Piccioni; an outwardly identical but lighter pale blue machine in group B for André Morel; and a diminutive 10CV *Petit Laboratoire* in Group C for Lefèbvre. All were built around aluminium and plywood monocoques along the same lines as the 1923 Tours cars, but without the ultra-narrow rear track that made the C6 so tricky to handle on the limit. The group A cars were fitted with a bombproof version of the high compression four-litre engine of the production C5 Sport, with a heftier crankshaft and bearings. Although the mechanical efficiency was inferior to that of the standard car's, these bespoke all-aluminium dry sump units were good for 130bhp at 3500rpm.

In 1961 Gabriel Voisin, modestly describing the C9 as the finest petrol-engined car of any type he had ever driven, scapegoated Rougier for their lack of success at Lyon – an especially bitter pill to swallow, since it left the field clear for a clean sweep by the despised Peugeots. 'My strict stipulations were subverted by the idiots I employed at the time,' he fumed. 'This chaotic state of affairs allowed undesirable features to be introduced, and our cars for the Grand Prix differed considerably in detail from their intended specification. This lack of discipline cost us victory, and I've never forgiven myself for my weakness in dealing with those who abused the trust I placed in them.'

All Rougier had in fact done was to arrange a lucrative deal on his own account with another carburetter company so that, unlike the others, his C9 was not equipped with the new Solex. 'Rougier lost 33 minutes at the start as a result of the carburetter he had imposed on us to earn himself a tidy sum,' railed Voisin. 'I was tearing my hair out. Having drained the batteries with his futile attempts to start the car, our so-called champion struggled breathlessly with the starting handle, to no effect. The engine was flooded, and I eventually had to spell out the need to remove the plugs and turn over the engine before replacing them. When the car eventually did start, Rougier was in a highly agitated state. He was no longer a young man, and the time he had lost and the penalties he had accrued by cranking the engine made him careless. He took risks during the night, and after entering a 90-degree bend much too fast he ended up in a ditch from which he could not extricate himself.'

Thus, in some ignominy and with a broken arm, ended Rougier's last competition drive. Exactly when he established his Paris agency for Steyr is unclear, but the combination of blotting his copybook with the famously irascible Voisin and the fact that he had competed at Lyon against the 4.4-litre Type VI Steyr streamlined saloons driven by Hermann Rützler and Ulrich Kinski must have been a factor. He managed Steyr's entry in the 1925 event at Montlhéry in which, despite being plagued by problems with the Michelin Comfort balloon tyres, Gaudermann finished second in the large car class.

Thereafter, Rougier concentrated on his dealerships and related business interests. In 1924, the Society of Automotive Engineers reported on 'the new Rougier plywood body', but more information is hard to find. In 1927 he lent his illustrious name to the manufacture of a flexible safety steering wheel in the manner of René Thomas, but of his own design. By 1936, Garage Rougier in the rue de Cormeille closed its doors for the last time. For nearly 30 years, the Marseille maverick had made a vigorous contribution to the early days of aeronautics and the automobile in general, and to Voisin in particular. Dapper, daring and determined, he added energy and colour to the modernity he so enthusiastically embraced, and deserves to be better remembered. He died in 1956.

The author thanks Robert Dick, Stefan Ittner, Francis Metzger and Philippe Ladure for their help with this article.

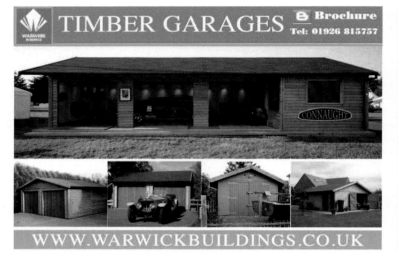

HIGHTONE RESTORATIONS
NORTH OXFORDSHIRE SPECIALISTS
HOPCROFTS HOLT, STEEPLE ASTON, BICESTER, OX25 5QQ

Complete restorations, servicing and repair of Vintage, Thoroughbred and Classic cars, including:
Panel-making; Accident repairs; Painting; Trimming; Ash framing; Re-wiring; Mechanical rebuilding and servicing.

— All work carried out to the highest standards —

Contact Hightone Restorations — tel.01869 349003.
Fax. 01869 349009. Email info@hightone.co.uk
Please visit our website - www.hightone.co.uk

FRENCHAY GARAGE
COACHBUILDERS & RESTORERS
* A complete in-house service *
* Vintage Alfa-Romeo parts always made to original pattern *
* Brakes and clutches relined *
* Ash frames, complete bodies *
* Engines overhauled *   Dinitrol rust preventative treatment available   Approved Stockist Castrol Classic Oils
* Cellulose painting *
* Rewiring & discreet Car Alarms fitted *

Frenchay Garage
(Incorporating Vintage Frictions)
Frenchay Common, Frenchay, Bristol BS16 1NB.
☎ (0117) 956 7303 ☎ (0117) 960 3708

If you want to sell your car
WE CAN HELP
We sell cars on consignment, efficiently, discreetly, reliably.
Why be satisfied with auction prices?
Why be bothered by time-wasters? Why be plagued with part-exchanges? Leave it all to long-established, successful professionals who will care for your car and market it with pride. We understand your type of car because we are enthusiasts ourselves and use such cars. Take advantage of our spacious premises, ideally located in North Oxfordshire, and our world-wide contacts **FREE OF CHARGE!**
LET US COLLECT, MARKET AND SELL YOUR CAR LIKE WE HAVE DONE FOR SO MANY SATISFIED CLIENTS
We need stock — now

Malcolm C. Elder & Son   Tel/ fax:01869 340999

STORAGE
BERKSHIRE. M4 J11 3 MILES

20 mins from Heathrow, choice of standard or dehumidified, Transport & Valeting, Service and MOT, very secure, owner lives on site

RICHARD THORNE

Tel No: 0118 983 1200
Fax No: 0118 983 1414

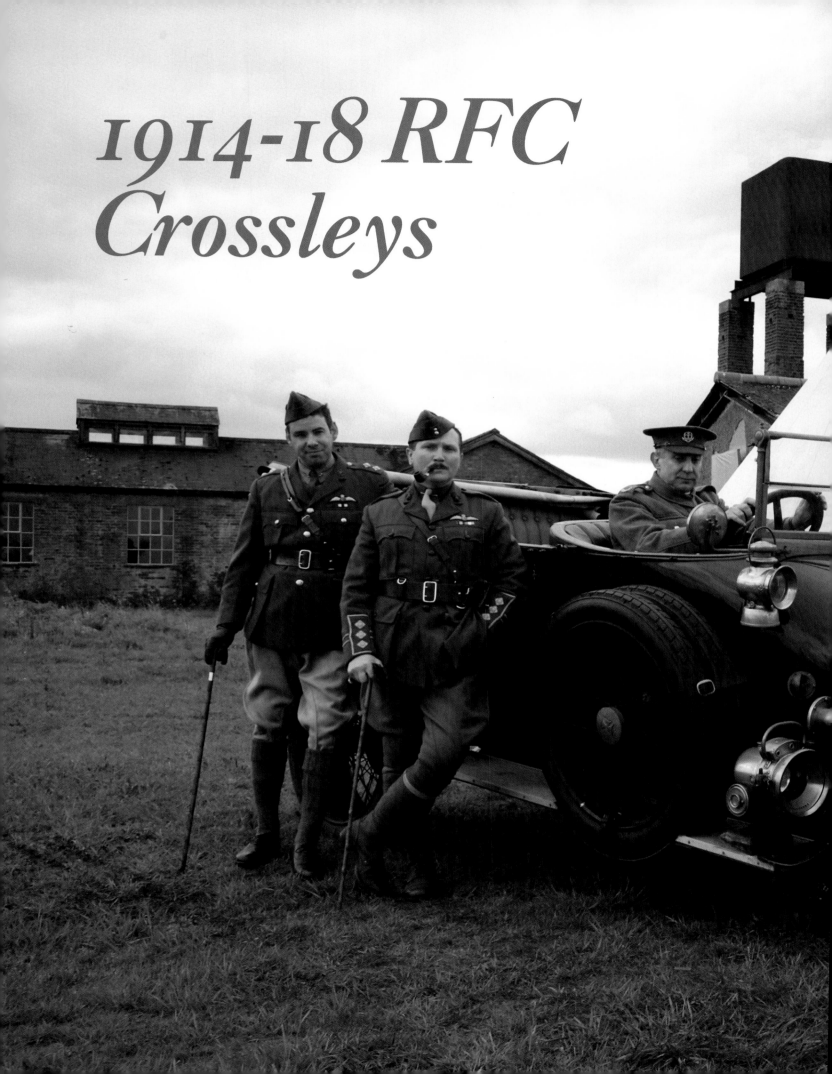

1914-18 RFC Crossleys

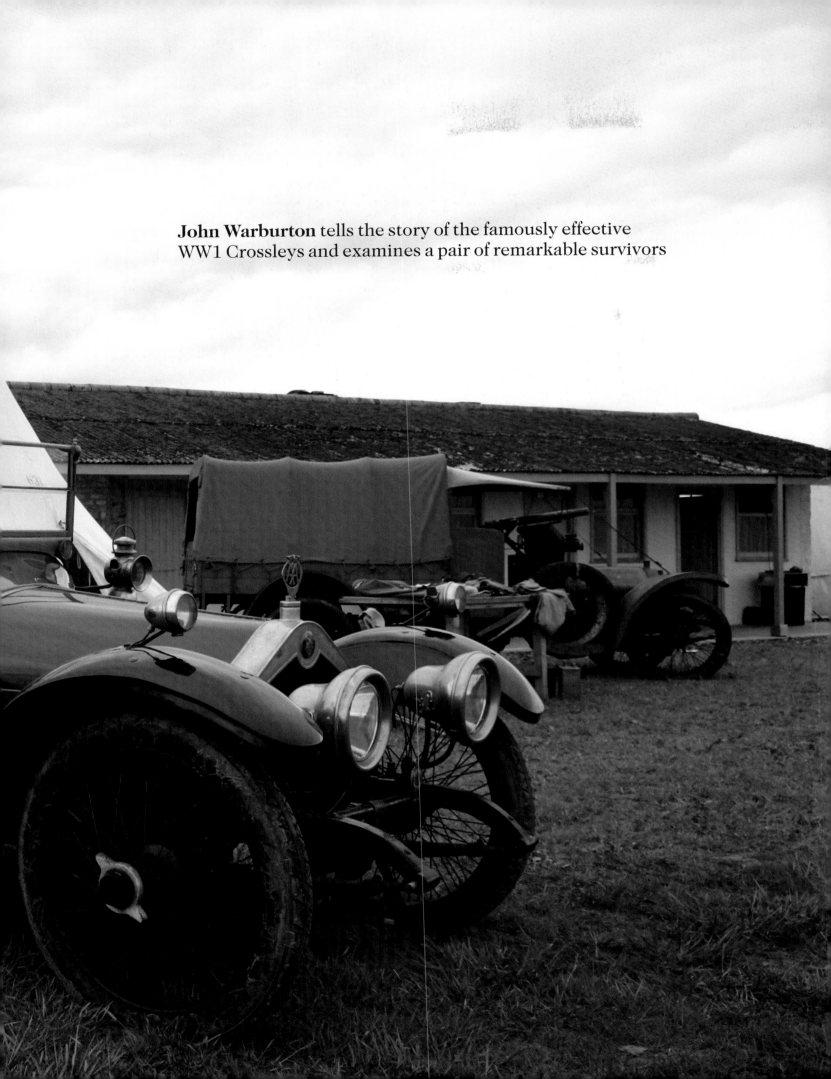

John Warburton tells the story of the famously effective WW1 Crossleys and examines a pair of remarkable survivors

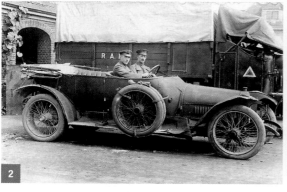

Pages 96-97. This wonderful photograph of Tom Fryars's two Crossleys was taken in May, 2012, at Stow Maries in Suffolk. Stow Maries is the last WW1 RFC airfield that survives intact in its original layout; more information can be found online at www.stowmaries.com
The men in the picture are members of a World War One re-enactment group, The Great War Society, whose website is at www.thegreatwarsociety.com
(*Stephen Le Grys*)

1. A typical Crossley tender on service during the First World War

2. A battered 20-25hp staff car in a military setting; the windscreen has been lost and much evidence of rough usage in evident

3. A well-filled 20-25hp staff car taking visitors round Snowdonia in the 1920s

A recent discovery came as a considerable surprise for those interested in the history of the Crossley car. The Directors' board meeting minute books of Crossley Brothers Ltd, held by the Museum of Science and Industry in Manchester, show that from late 1913 until well into 1914 the Directors were discussing phasing out car manufacture altogether.

One can speculate upon the possible reasons for this; financial considerations surely were paramount. Whilst the two models in production, the 15hp and the 20-25hp, were on a par from aspects of design, workmanship and finish with market place rivals, this medium-size 'quality car' sector was amply supplied. Wolseley, Sunbeam, Vauxhall and Daimler all spring readily to mind, even without the popular imported makes being considered. It is likely that, in their attempts to be competitive, Crossley Motors Ltd, the car-making subsidiary of Crossley Brothers Ltd, had been forced by stiff competition to price their chassis at figures that minimised profits. The records show recurring cash injections from the parent company, and an order for 25 20-25hp War Office cars early in 1914 called for a reluctant advance of 'another £5000'. But this order was to be the providential light at the end of the tunnel. The outbreak of war was to reverse a dire situation where the need to liquidate the subsidiary company had appeared on the agenda of each monthly Directors' meeting.

Earlier, with the clouds of war looming, the Government had authorised the Army to hold trials for vehicles suitable for military use. Crossleys entered their vehicles, and their robust qualities and reliability had impressed the authorities. This then led to that order for the 25 20-25hp chassis, and within weeks the declaration of war on 4th August saw an urgent need for many more such vehicles.

Unlike many of their rivals, Crossleys had the facilities and workforce to enable them to deliver large numbers of staff cars, light tenders, ambulances and similarly-bodied chassis for the war effort. The Forces would undertake to have all the chassis that the firm could build; at this time, most other car makers faced the need to abandon motor manufacture and seek contracts for munitions, military equipment, aircraft components and suchlike products, for which supplies of raw materials would be approved. Few were the potential buyers of new cars in any case at such a time of national crisis. For Crossley Motors, not only did this arrangement continue through to the Armistice of November, 1918, but smaller orders, principally for

Howard Hughes took his enthusiasms seriously – film, flying and cars, particularly steam cars. Great wealth flowed from Hughes Tool, manufacturing and leasing the drill bits which produced the world's petroleum, and was spent with style. His first big Hollywood effort was *Hells Angels*, still a cult flying movie, which cost $3.8 million in 1930. *Scarface* and *Outlaw* were both critical and commercial successes despite (or perhaps because of) problems with the censors, the first thought too violent and *Outlaw* on account of the generous display of Jane Russell's attractions.

Flying was a more serious matter. The Hughes Racer was an advanced design, beautifully built, which set the world landplane and trans-USA speed records. Next, Hughes took the round-the-world record in a comfortable 91 hours flying a Lockheed Vega. As a pilot he combined skill with real personal courage, surviving two near fatal crashes in experimental aircraft. This was not a playboy hobby: a 1939 investment of $7m in TWA turned into a profit of $547m on its forced sale in 1966. Hughes Aircraft, left to the charitable Howard Hughes Medical Institute, was eventually sold to Boeing for $5.2bn. The final effort was the *Spruce Goose*. The nickname which Hughes hated was applied to a troop-carrying flying boat with 320ft wingspan (a 747 spans 195ft) and 24,000hp from its eight 28-cylinder Pratt and Whitney R-4360 radials. It flew once, for one mile with Hughes at the controls, and is preserved in Oregon.

And then there were cars: a billionaire's hobby, but always with a serious purpose. As a young man about Hollywood, the backbone of his stable was the standard millionaire's luxury fleet – Cadillacs, Packards, Minervas and four Silver Ghosts, one of which he gave to then-current girlfriend Gloria Swanson. He became the only customer to buy two new steam cars from Abner Doble. These he retained, uprated and fitted with sponsored improvements. Nothing in his stable could out-run E22, the faster of the Dobles, until much later a Model J Duesenberg appeared. Significantly, in 1929 he also developed the Hughes steam car, built on a front-wheel-drive IFS chassis said to be of French origin and fitted with a poppet-valve swashplate engine. The car was not a success, probably through poor steam raising rather than mechanical difficulties, but the power unit with its integral brakes and front-drive was in many ways an advance in steam practice.

Hughes's admirable attitude to life lives on. Today's two-Doble owner lives in Yorkshire and it is one of his cars, the ex-Hughes E22, which *The Automobile* was invited to inspect. Barry Herbert came to steam cars via three Stanleys, including a 1911 Model 85, an example of the production American steam car at its Edwardian best. The Stanley twins epitomised the New England Yankee: utterly honest, occasionally bloody-minded, humane, and first-class businessmen. They made their money by selling a successful photographic patent to Kodak and expanded their steam car business as much for fun as profit. The Stanley was simple, light, strong enough to withstand New England roads and easy to drive. Within two years they had met orders for more than 200 cars and sold the business, very profitably and more or less by mistake. Having relaunched it as Locomobile, the new owners shortly decided that the future lay in petrol cars and sold the steam business back to the Stanleys, who found themselves with an extended factory, a profitable patent deal and $250,000 to play with.

By 1905 the twins were building the definitive Edwardian Stanley with a simple vertical fire tube boiler and petrol-fired vapourising burner under a round-nosed bonnet. The nose of the twin-cylinder slidevalve horizontal engine was pivoted on a crossframe, and the crank was integrated with the differential; the engine was reversible, no gearbox was needed and unsprung weight didn't much matter. Start-up was easier after 1906 when Prestolite acetylene became available for preheating the burner, but could still take half an hour. The other snag was high water consumption with the non-condensing unit. The twins took a philosophical view. If you wanted to leave early, you got up earlier. F O Stanley is alleged to have told a customer critical of the short range, "But, my dear sir, 30 miles is all you will ever want to drive in one day."

An endearing feature was that bumping up

1. Howard Hughes, photographed in the 1940s in front of a Boeing 100A in Inglewood, California, enjoyed fast cars as well as aeroplanes

2. The mighty ex-Hughes Doble steamer, E22: beautifully proportioned, incredibly complex and very, very expensive when new

boiler size and pressure could make Auntie pick up her respectable skirts and run. This was taken to extremes in Fred Marriott's Daytona car, which reached 127mph in 1906. Next year was a record too far: the car was travelling at about 150mph when it flipped. Marriott survived. The company retired from competition but hedged its bets by building a Gentleman's Speedy Roadster capable of at least 75mph in standard form. Stanleys are still useful performers, but the real attraction is the manner of their going: Stanley motoring on a smooth road is an enchanting experience, with the wind more apparent than any mechanical noise. Suspension is smooth and progress brings an uncanny sense of detachment which even a Silver Ghost doesn't quite match. In production terms, Stanley was the most successful steam car manufacturer. The only problem today, unlikely to have been a serious issue in America before the Kaiser war, is the deplorable brakes.

Their robust design philosophy lasted to 1916, when pressed steel replaced the wood frame and semi-elliptics appeared at the front but not at the rear, where perch rods still located full elliptics. Front-wheel brakes never featured. At last a condenser cut water consumption and eliminated the characteristic vapour trail, but it brought problems of its own through water contamination with engine oil. The brothers retired in 1917. Excess weight sapped the performance of the later cars, and by 1923 the original firm was bankrupt. Its successor folded in 1926.

Abner Doble met the Stanleys as a teenager and was sufficiently impressed to start production of his own steam car. With real prescience he identified the two main drawbacks to the Edwardian Stanley: it took too long to start and didn't condense its exhaust steam. By 1912 the 17-year-old had built seven cars, keeping the final Model A, whose outstanding feature was a condenser with a filtration system that avoided oil contamination in the boiler. A Model B was followed by the controversial Doble-Detroit, launched to massive acclaim just as the US Government introduced war emergency restrictions. The innovative uniflow engine seems to have been troublesome and only 11 cars were completed.

By now Doble had established his strengths and weaknesses. He had appreciated the enormous appeal of the steam car, identified the drawbacks of existing models and proved his ability to build something demonstrably better which could be sold at a premium price in a market dominated by the petrol car. Like Henry Royce, he had an insatiable appetite for hard work and fast development; the downside, as with Royce, was that his perfectionism led to an inability to leave well alone, coupled with a certain disregard for cost. Of the 32 cars built post-war no two are alike, and a few engineering snags persisted to the end. Barry Herbert, a very practical perfectionist, has addressed some of these snags, using technology not available to Doble, and has demonstrated results of which Abner could be proud.

Herbert's Doble epiphany came when his Stanley 85 was joined by Doble E11 during an event in Australia. He describes the encounter on the first page of his book, *A Tale of Two Dobles*, reviewed in *The Automobile* in June, 2011, confessing 'I immediately fell in love with it.' It turned out to be a long, expensive but ultimately happy affair. A search led to his buying, with some difficulty, serial D2, only survivor of four 'development chassis' built in 1921 when Abner Doble was still refining the specification of his post-war supercar. No documentation existed, and Herbert pays generous tribute to the friends and colleagues who helped with the work.

Doble D2 turned out to be something of a hybrid, with the later E-type steam generator. The chassis, engine and complex auxiliaries appeared in decent order, although electrics were a problem. Replacement of a non-original body with a four-door tourer turned out to match the original Murphy design: the new coachwork by Michael Riley of Derby looks superb. The engine is a straightforward compound twin. Sorting out valvegear installed upside down was only a problem on account of lack of accessibility, a chronic Doble design issue. Predictably, the steam generator and its auxiliaries were difficult, but the real problems came with the condenser system. A redesign partially works

3. Gigantic headlamps dominate the frontal aspect of this enormous car. The 'OIL' number plate has been added by the current owner

4. Not a name you come across every day... Doble steamers are pretty difficult to find anywhere, but especially so in the UK

5. The polished dashboard looks conventional enough until you analyse the functions of the numerous instruments

but Herbert writes, sadly and realistically, 'Having come to the conclusion that D2, being a development chassis, had never condensed, I do not intend to spend any more time on the condensing situation.' This and other issues, including inadequate braking, were resolved eventually, however, and D2 performs beautifully.

Appetite, as some gluttonous Frenchman pointed out, grows with eating. With perhaps the most tricky of Doble restorations complete, why not another one? Differences in specification might mean that the learning curve would start all over again, but at least Herbert knew the problems. After a brisk pursuit, E22, Howard Hughes's second Doble, joined the Yorkshire stable from a private owner in the Channel Islands. The car was one of a run of 20 E-types, 12 surviving. This was series production by Doble standards, despite variations in each. One might hope for fewer unknowns than in the lone survivor of the experimental D-class. The fact that the E-type was largely in boxes with the bits unlabelled was one of those things: at least E22 retained its original two-door body by Murphy of Pasadena, with the very American feature of a transverse golf club locker.

E22 represented Doble's mature thinking about a powerful and luxurious steam car. The massive nickel chrome steel chassis was specially lengthened during an early factory rebuild to give a 150in wheelbase, making room for updates to Hughes's specification including a 30 per cent deeper condensing radiator and a turbo-booster. This equivalent of a petrol engine turbo uses exhaust steam to deliver air at greater than atmospheric pressure to the combustion chamber, increasing the intensity of combustion and steam output. Turbo characteristics are self-compensating, increasing output to match demand. The result was 100mph-plus performance in a 2½-ton car, an interesting thought given that the front-wheel brakes were a 1928 retrofit. The price was $20,000. Inflation and the fall of sterling against the dollar make comparisons difficult, but in today's money Hughes paid well over £1,000,000 for his new toy.

The engine of E22 is a four-cylinder double-acting compound. In other words, steam is fed by piston valves first to the high pressure cylinders and then transfers for further expansion in the low pressure pair, resulting in better (although still by IC standards poor) thermal efficiency. The engine is integral with the rear axle, making a unit of impressive size and weight. A separate auxiliary unit driven from the main engine houses four water pumps, vacuum pumps and a hefty dynamo. Engine output is a useful 125bhp at normal boiler pressure, with a frightening 1000lb ft torque at zero rpm. Main castings were in good order but a new crank from Brineton, suppliers to Bugattisti, new roller main bearings and piston valves were needed. Yet again, valvegear had to be reassembled the right way round.

The original steam generator is the flash type, with water pumped through more than 550 feet of steel tube tested to 7000psi. Working pressure is 750psi. and the control system includes a normaliser which manages feed water flow and is key to successful running. After the problems with D2, Herbert looked inside the combustion chamber of E22, didn't like what he saw and commissioned a new one. He then considered the steam generator and called Haycock and Hague in Hull. They were quite used to Dobles by now and their new unit works well. Accessories, however, were more complicated. A surprising amount of electrical power is needed: ease of driving and engine control come at the expense of elaborate solenoid systems which have to be just right, and more power is needed for the electric motor driving the blower motor. The D2 with an electric blower motor had a maximum demand of 100 amps. Happily, this was not required on E22 and Herbert, very rationally, solved his problems with an alternator. Cooling air is helped through the condenser by a four-bladed 24in diameter fan, with its private turbine driven at 3000rpm by exhaust steam. The 400 bronze turbine blades riveted to an alloy main casting were given a clean bill of health by a turbine manufacturer, which is just as well: explosive failures have happened to others and would mean a £20k bill.

The turbo-booster and burner were challenging. E22 was one of four E model cars to start life with a turbo-booster, which at some point was removed and fitted to another car. The brave decision was taken not simply to remake this, but to replace it with a new F-type booster, a period retrofit on several cars. Unlike the neat little bolt-on turbo on a petrol car, this is a massive cast assembly housing a two-stage system with a second fan reaching 8000rpm. On the road it makes the difference between 80 and 120 mph. An electric blower, using a 3300rpm motor drawing almost 24 amps, gave the restoration team breathing space before a scheduled trip to Australia. It also provided extra time to attend to the often temperamental burner. E22 was tried briefly by Doble with a Hughes-sponsored forced-draught vapourising burner but the standard unit was later refitted. This relies on a simple float carburetter fed from the blower and supplying the combustion chamber above the steam generator coil section. The vital control box is mounted on the side of the generator, and exhaust gas is expelled through a large duct at the base of the unit. A drain pipe from the carburetter handles any possible overflow: experience showed that it is important for this to run clear of the car, so that any minor fire is safely out of the way.

Work had already started on E22's original body at Michael Riley's Derby establishment when Herbert acquired the car. This continued with input from the new owner who said, most rationally, 'I wasn't going for a Pebble Beach finish, just a highly desirable road car.' This and more was achieved: all 18½ feet of E22 looks impressive in grey with black wings, the correct R-R Springfield pattern lights, restrained chrome trim and Australian buffalo hide. Speed of progress is even more impressive. The learning curve of Doble D2 must have helped, but the timescale is startling to anyone who has ever had to wait for action on a long, drawn-out restoration project. E22 was acquired in May, 2006, and delivered to the owner for testing and commissioning in January, 2009. At the end of the month it was loaded into a container for Australia to take part in the Outback Vintage event. Naturally, there were problems: final testing in the UK resolved relatively minor issues, while others including heavy steering and poor brakes showed up during another 500 miles in Australia. One minor fire was put out. Back in the UK the steering was rectified and turned out to be pleasantly light and very precise. Changes to the brake hydraulics, and fitting a 12-volt vacuum servo, made the heavy car manageable and safe in modern traffic.

Modifications to the steam generator and combustion chamber were needed to eliminate hotspots. Changes to the valve timing – in steam terms, notching up – produced full power and a water consumption of about 120 miles on a tank, a far cry from Stanley's 30 miles on a good day. Unfortunately, the combustion arrangements of the Doble have always been tricky. A recent fire caused by buildup under the car of fumes from the carburetter damaged the complex wiring but, happily, little else, leading to a decision to rewire and to convert to a pressure jet system burning a less volatile 50/50 petrol/diesel mix. This work is again being carried out by Riley in Derby, where we were able to take photographs. We greatly look forward to seeing E22 back on the road.

Abner Doble continued to provide consultancy to other companies, including Sentinel in the UK and Henschel in Germany, but Doble Steam Motors Inc of Emeryville, Oakland, California, was the last to build steam passenger cars on a commercial scale.

A direct comparison with equivalent cars in the Hughes stable is possible. Doble E22 in its present form, without a turbo-compressor, weighs about 2½ tons, uses water at a rate of 6mpg and has a fuel consumption of roughly 7-8mpg. Herbert reports an easy 50mph on Australian roads, with perhaps 30mph more to come. As we know, Hughes paid $20,000 for the car. His Springfield Rolls-Royce Silver Ghost would have cost rather less: in 1924 the bare chassis was priced at $11,385, the cheapest Pall Mall tourer $12,930 and the Mayfair Town Car $15,880. Special fittings of the sort demanded by Hughes would have closed the price gap; however, as current supercars have shown, this rarefied market is not particularly price sensitive. More economical options included a Cadillac V-8

Supreme Steamer

Imperial Limousine at $4400 or the excellent V-12 Packard for less than $7000. A Model T roadster cost $260. The difference in performance between the Doble and the much lighter Ghost would be great, although the Royce would have used rather less petrol and incomparably less water. Neither consideration is likely to have worried Hughes much; he loved the potential 120mph top speed of the Doble and its fantastic acceleration.

The Doble steam car had illustrious predecessors. The Veteran Stanley is an infinitely more civilised means of transport on a fine day than is afforded via the bum-numbing vibration of a comparable single-cylinder petrol voiturette. Edwardian Stanleys still invite magic carpet clichés. In France, the Serpollet was a serious challenger in the European luxury market. None were really market contenders after the Great War, but interest in steam on the roads continues.

The old claim that steam cars were killed off by oil company interests doesn't stand up, as even by the 1920s the internal combustion engine used less fuel than the steamer. The main steam contenders had died (Serpollet), given up steam cars (White) or just plain given up (Stanley), at the very point where the petrol car had become a practical machine rather than a rich person's toy. Steam car development from 1915 onwards was largely in the hands of small-time entrepreneurs, while the mainstream motor manufacturers had no need of a technical diversion from the petrol engine which did the job and which they knew how to make and sell. While a simple steam car such as the commercially successful Stanley provides a knowledgeable owner with delightful transport, the controls needed to make it appeal in the mass market are too expensive and complex. Today the universal silicon chip and ECU could handle steam as well as petrol and the steam car's burner, operating at low pressure and relatively low temperature, can more easily achieve clean combustion than the internal combustion engine with no need for catalytic converter or particulate filters. Clouds of steam, so characteristic of the early steam cars, had already been eliminated by a radiator/condenser to recycle it.

More recent work includes the Paxton Phoenix, financed by chainsaw profits and with Doble as a consultant, a serious contender which never reached production. Lear, of LearJet fame, also spent substantial sums. General Motors has admitted 'feasibility studies', Issigonis worked on a steam-powered Mini and even Saab had a steam project, but none of these progressed beyond the prototype stage. Although other industry majors have worked on steam cars quite recently, these are probably, metaphorically at least, on the back burner. The modern IC engine is highly fuel-efficient, largely through high compression ratios and small heat losses. The traditional form of steam engine cannot compete in efficiency and economy and so suffers higher fuel consumption and CO_2 emissions; neither of these shortcomings is socially, politically or commercially acceptable. 'Steam' engines can be more attractive running on fluids other than water at very high pressure, but the engineering and, particularly, metallurgical problems are severe and the alternative working fluid seems to have been abandoned. The trend in modern transport to smaller packages would call for a rethink of the fundamentals of external combustion.

Yet the modern steam engine continues to evolve and may yet power a car. Cyclone Power Technologies Inc, based in Florida, is developing a compact power unit which operates with supercritical steam at 3000psi, a pressure at which water and steam have the same density. Using a closed circuit comprising burner, coiled steam generator, a six-cylinder radial engine and a centrifugal condenser, all the moving parts are lubricated by water rather than oil, which neatly solves the problem of separating oil from water in a closed-circuit steam power unit. It remains to be seen whether this machine will become a commercial proposition, but we should hear more of it soon as it is hoped to use one to regain the Steam Land Speed Record for the USA.

The author is most grateful for help and advice from acknowledged steam car experts Barry Herbert and Mike Clark.

1. Doble E22 was more expensive than a Rolls-Royce but, unlike any Rolls of its time, used even more water than fuel

2. The present owner has made numerous detail improvements under the bonnet, enabling the car to run on a less volatile fuel mix

3. Owner Barry Herbert has added this visible reminder of Howard Hughes's ownership of this particular car

4. The chassis plate gives its date of manufacture as 6th September, 1925

5. E22's beautifully finished coachwork is by Murphy, of Pasadena

07719995514 (M)
Tel: 01564 779746 (H)
PX A PLEASURE – USUALLY – Nr Junction 5, M42
Available 7 days a week 9am - 10pm — Sad isn't it!

Bob's Affordable Classics

MGYA 1951
This is quite possibly the best car of this model on offer anywhere. Excellent structure, no rot, paintwork is blue body with black wings, all seats are retrimmed in leather and the woodwork is excellent. New Avon tyres, the chrome is excellent and the car has been rewired. It recently passed its MOT, ready to enjoy.
£17,500

BUICK MCLAUGHLIN 8 1938 90 SERIES LIMITED
Imported new in 1938 - right hand drive. Believed to be the only one in the UK. One registered keeper for 65 years. The car can seat 8 people, glass partition. The bright work is bright and the paint work is Deep Blue metallic. Totally retrimmed and painted. This magnificent car is a joy to behold - now ready for work or pleasure. As much as it hurts me I could be parted from this beautifully buxom and sexy lady for: **£25,000**

FERRARI 456 GTA AUTOMATIC
1996. Metallic blue with contrasting grey leather interior. Only 25,000 miles with service history. Four stamps and two cambelt changes. Recent cambelt. Alarmed and Tracker fitted.
£29,950

BENTLEY TURBO R LWB
1997, so one of the last. Cherished reg included – WJW 214. Lots of options – wood steering wheel alone was a £1600 extra. Excellent service history.
£12,950

ROVER P4
1954, the engine is from a P4 110 with Westlake head, overdrive gearbox, low ratio back axle and uprated brakes. Therefore, the car will cruise comfortably at motorway speed. The cellulose paintwork is exceptionally good and the seats have been re-trimmed in period spec. **£8,950**

JAGUAR X300
6 litre long wheel base, originally owned by Jaguar. Last owner is a doctor of engineering and top jolly at Rolls Royce. Excellent service history. Superb condition, silver and grey leather interior. **£5,950**

TRANS AM TURBO
1981, very bright and tidy, alloys, low mileage, taxed & MoT'd, the bird on the bonnet is a bit flakey round the edges, but then all my girls are, it's probably my age! **£7,950**

LANCIA INTEGRALE 8V 4WD TURBO 1988
Power steering, Electric windows, Momo steering wheel, Alarm and immobiliser. 81,000 miles (130k Km) Service history and recently had a new cam belt and water pump and 2 tyres. **£5,995**

MINI 1000cc
30th anniversary model – de-seemed, retro door hinges, flared arches, special exhaust system, alloy rear roll cage (this can be removed if you wish) Cooper bumpers. Sound system, DVD player. The colour is a superb bright orange – colour coded Superlite wide alloy wheels. All this and only 19,000 miles from new. **£7,950**

BIG DIAMOND RING

Sorry, I have run out of talent!
This is a rubbish photo of a most beautiful diamond ring, set in platinum.
Approx. 3 carats - stuck on a carrot!
Valued last year at approx. £40k
(the ring, not the carrot).

We are selling now as it does not look at its' best on a finger riddled with arthritis and liver spots and the chance of me finding anyone with younger, prettier hands is nil!

Will do a deal on a car, plus cash, to the value of £20,000

I'm available 7 days a week 9am - 10pm — sad isn't it!!!

1906 Grand Prix Darracq

Built for and run at the first French Grand Prix at LeMans driven by Louis Wagner, this car went on to win the Vanderbilt Cup in America on the 6th October 1906. It was later purchased by Malcom Campbell who campaigned it at Brooklands from 1910 to 1913 and was the first of his vehicles named Bluebird. Onsold in 1913 to Neville Minchin it was fitted with a touring body, later going to Ireland where the car was damaged by fire. An extensive rebuild was completed in 2005 and in 2006 the car was the centrepiece of the ACF's Centenaire celebrations at LeMans. It was also campaigned with great success at a number of VSCC events including winning Fastest Edwardian Climb at Prescott driven by owner Anne Thomson.

This 100+ MPH car is now offered for sale, for further information contact Anne by email at anne@anne.co.nz

STANLEY MANN RACING

1929 BENTLEY 4½ LITRE LE MANS SPEC
MATCHING NUMBERS
REGISTRATION NUMBER: EC 8504   CHASSIS NUMBER: FB 3304   ENGINE NUMBER: FB 3304

All components to this car are original so she has the much desired MATCHING NUMBERS.
A stunning restoration on this BENTLEY 4½LTR LE MANS shows a fine quality of workmanship rarely surpassed..
Finished in British Racing Green with matching green leather.
Drives extremely well and would make a great rally or touring stead with excellent performance to boot

WWW.STANLEYMANN.COM   Tel: +44 (0)1923 852 505

Carburetters to Compressors:
René Cozette's Contribution

France's many sports and racing cars of the 1920s gained power and speed from the novel compressors designed and built by René Cozette. **Karl Ludvigsen** describes a passion that reached to America and the world's largest car manufacturer, General Motors

When supercharging for automobiles burst on the scene after the Great War, French manufacturers and enthusiasts had an ample domestic choice of suppliers of proprietary compressors. In the vanguard was carburetter manufacturer René Jean Paul Emile Cozette, whose established speciality alerted him early to the coming demand for automotive compressors.

Among the world's independent entrepreneurs and actors in the field of the supercharger in the mid-1920s Cozette had few rivals. Born in 1895, he received his first patents on engine sealing in 1914. During the Great War he tended the engines of France's aviators as mechanic and technician, gaining the knowledge that led to his first carburetter patent in 1919. Following with dozens more, Cozette then applied for his first supercharger patent on 28th November, 1924. In this discipline as well he would act to protect his improvements over half a decade.

Cozette returned to Paris in 1924 after a sojourn in America, to which he was invited by spark plug maker Albert Champion. He added superchargers to his existing products, which included carburetters as well as high-compression cylinder heads on principles licensed to him by Harry Ricardo. Cozette was the designer behind the sleeve-valve engines that racing driver Albert Guyot used in cars of his own name that competed in Grands Prix, albeit with meagre success.

René Cozette adopted the vane-type compressor, which as its name implies wipes a crescent-shaped chamber with a series of vanes to achieve internal compression of gases before releasing them to the engine. To keep the tips of the vanes from rubbing against the housing he inserted a thin rotating drum, dubbed the 'false rotor', between those tips and the housing proper. This recalled a method used by William Reavell in England, but with the important difference that Cozette positively controlled a false rotor that Reavell allowed to turn freely. Cozette synchronised its rotation with that of the vane-carrying rotor in a lock step so that the vane tips slid back and forth, during rotation, on the false rotor's inner surface.

Although a degree of rubbing between the vanes and the false rotor still prevailed, this was far less than would ordinarily occur. In the words of another supercharger inventor, when such a 'cylindrical rotating element' is used 'the linear friction of the blade, instead of being equal to the periphery of the inside of the case, is only equal to approximately four times the eccentricity of the rotor with the case.' With the edges of the blades or vanes sliding back and forth over constrained areas, between those areas of the false rotor were the vital apertures through which the charge was drawn in, compressed and delivered. Cozette provided light springs to hold the vanes against the false rotor when centrifugal force alone was inadequate to effect a running seal.

Initially, he used a pair of gears to co-ordinate the motion of the two rotors. This turned out to be 'complicated and noisy', so he soon patented an arrangement that was 'kinematically much more agreeable'. A disc on the blower's driving shaft carried six small rollers at its periphery. These rollers engaged with six circular recesses in another disc that was screwed, at its outer edge, to the false rotor. The shaft and rollers simply carried the false rotor with them. Though ingeniously simple in concept, its implementation demanded a multitude of small parts and screws that were unique to the Cozette supercharger.

In December of 1925 Cozette filed for a patent on an application that provided two bypasses between the carburetter, at the blower inlet, and the delivery to the engine. The larger of the two, normally blocked by a butterfly, could be opened if the blower seized in such a way as to obstruct the flow, which would be disastrous if the engine were powering an aeroplane. The smaller bypass was designed to improve low-speed operation by reducing the blower's internal vacuum, which drew in excessive oil, and by dispersing the condensation of petrol in the charger.

René Cozette soon catalogued 11 models of compressor to suit engines of 250, 350, 500, 750, 1000, 1500, 2000, 2500 and 3000cc, all

COZETTE

CARBURATEURS
COMPRESSEURS
CULASSES

SOCIÉTÉ ANONYME DU

CARBURATEUR COZETTE

CAPITAL 3.000.000 DE FRANCS

34, Avenue du Roule -:- NEUILLY (Seine)

TÉLÉPHONE : GALVANI 83-77 et 93-61   R.C. Seine 206.966

1. René Cozette was the engineer behind the Guyot Special, here at Indy in 1926.

2. The Guyot Special had a two-stroke six-cylinder engine with sleeve valves, scavenged by a Cozette compressor seen alongside the engine. The SU carburetter is not original

3. A contemporary illustration showed the components of the Cozette: from left the housing, the two rotors, the synchronising mechanism and and end cover

supplied complete with his carburetter. The displacement per revolution of the blower ranged between 75 and 83 per cent of the capacity of the engine for which it was intended. In 1925 a Cozette blower helped a 1.1-litre Salmson set a new speed record for cyclecars at 113.1mph. Mounted on the four-cylinder engine's left side and driven by a train of gears from the rear of the crankshaft, it displaced the exhaust manifolding to the right of the bonnet.

In the first Cozette installation for Salmson the blower raised peak power to 58bhp at 4500rpm on a boost of 0.78 bar from the standard output of 33bhp at 4000rpm, the power needed to drive the compressor approaching seven bhp. In the San Sebastian evolution of the overhead-valve engine of Salmson's Grand Sport Course model the unblown output rose to 55bhp at 4000rpm. Fitting a Cozette, as masterminded by chief Salmson designer Jean-Émile Petit, raised this to as much as 70bhp. In 1926 a racing car of this specification was catalogued for sale by Salmson; both No 7 and No 8 Cozettes were offered.

For his works Salmsons in 1926 Petit fitted the No 8 Cozette for racing and the larger No 9 for short-distance sprints. While the Salmson four drove its twin overhead camshafts by a vertical king shaft at the front, Petit powered his compressors by a gear train from the crankshaft's clutch end. Mounted on the left, the Cozette was driven through a rubber coupling to isolate it from the engine's torsional pulsations. The latter were also dealt with by enlarging the diameter of the crankshaft within its rearmost bearing, a step by Petit that also acknowledged the extra power that was needed to drive the Cozette.

In the mid-1920s a Cozette was fitted to a four-cylinder 4.1-litre commercial-vehicle engine. Normally aspirated, its output was 68bhp at 2000rpm. With a drive ratio of 1.07 to one the compressor elevated this to 91bhp at 2300rpm with no change in the unit's 5.0 to one compression ratio. Though only a slight increase to 94bhp at 2200rpm resulted from a faster drive ratio of 1.42 to one, the power curve was much fatter over its whole range. Not surprisingly 'acceleration and hill-climbing were greatly improved.' On the debit side 'both petrol and oil consumption were increased, but the latter at any rate could doubtless have been avoided to some extent after further experiment.'

Carburetters to Compressors: René Cozette's Contribution

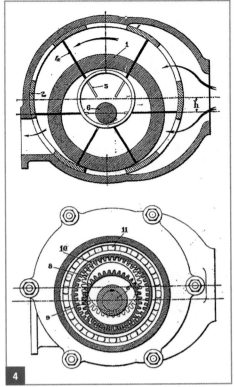
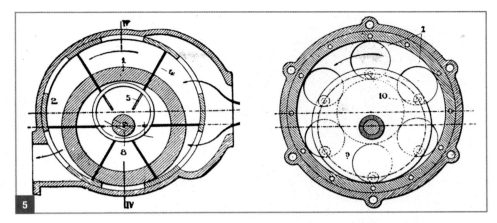
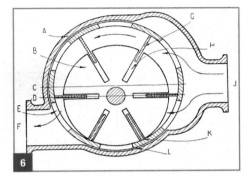
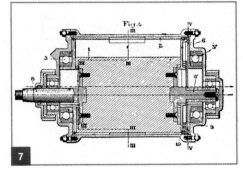

Soon chargers by Cozette were widely and successfully used by many car manufacturers in France and abroad. Prolific producer of proprietary engines SCAP standardised a Cozette-blown 1.1-litre four, which was fitted by Genestin and Tracta among others. Another SCAP customer was Bollack, Netter & Cie of Levallois-Perret, producer of BNC sports cars. It fitted a variety of available engines to its vehicles, wrote Tom Threlfall, 'to suit the capacity classes in the various events or possibly to suit variations in BNC's credit with the engine makers.'

In 1925 BNC offered a SCAP-engined sports car with a Cozette blower, the first supercharged model to be offered for sale to the general public in France. A 1.1-litre SCAP four powered BNC's Montlhéry model of 1927, which reached 85mph with the help of a Number 7 Cozette compressor, one of which had been tested on a works racer that first competed in 1926. Mounted vertically at the front behind BNC's sloping horseshoe-shaped radiator, the blower increased horsepower from 30 to 53. Its induction pipe was on the left, feeding a manifold topped by a pressure-relief valve. In stripped form the 1929 version of BNC's Montlhéry was guaranteed to better

4. René Cozette's original design used gearing at the end of the rotor to keep the rotation of the false rotor synchronised with the movement of the rotor itself

5. In a later design Cozette used an ingenious set of six rollers in circular apertures to keep the two rotors in synchronisation

6. Depicting the basic principle of the Cozette blower, which had a rotating 'false rotor' that held the tips of the vanes away from the outer housing

7. A longitudinal section of the Cozette blower showed the bearings carrying the two rotors and, at the right, the mechanism keeping them in lock step

100mph. Having taken over exotic sporting minnow Lombard, BNC used its 1.1-litre twin-cam six and Cozette-charged it to give 70bhp at 5500rpm.

Instead of Cozette's system of a separate lubricant tank with output controlled according to throttle position, BNC tapped the engine's oil supply. Its warmth was a boon to lubrication in cold weather, the company claimed. As well, it wanted to avoid excessive oil being sucked into the blower when the throttle was closed, resulting in spark-plug fouling. In BNC's patented system the engine oil still entered the Cozette through its fitting at the top, delivery controlled by a regulator screw. Also adjustable was the stop behind a spring-loaded ball valve that intervened in the delivery. At rest the ball blocked flow. It gave way when the engine started, admitting lubricant to the charger. But when the throttle closed, the resulting depression inside the Cozette sucked the ball against its orifice, overcoming spring pressure and cutting off the oil supply.

Another customer for proprietary engines

1. On the bench at compressor specialist Derek Chinn, the components of a Cozette, showing the ports in the false rotor that were essential to allow the gases to flow in and out of the blower

2. SCAP offered Cozette-blown engines, with the charger mounted vertically as Cozette preferred in order to provide the best inlet ducting

3. Installed here in a Rally, the 1.1-litre Type K Ruby four was expressly designed for racing with a Cozette blower. The engine was also used by Caban and D'Yrsan

4. On a special 16-valve model intended for competition, the two-litre Sizaire Frères mounted a Cozette behind the engine. It did not proceed to production

was Caban, the marque founded in 1926 by racing driver Yves Giraud-Cabantous. A neighbour of its Levallois-Perret works was H Godefroy et Lévêque, makers of the Ruby engine. Cabans were powered by various 1.1-litre Ruby fours with pushrod-operated overhead valves. For racing versions in 1931 Giraud-Cabantous used the Type K or 'Kappa' Ruby, sometimes fitting a Cozette-blown version to elevate output from 43 to 60bhp. Even against the ruling Salmsons these Cabans achieved some class successes.

Ruby's overhead-valve Type K was a serious racing engine with three main bearings instead of the usual two in a massive aluminium crankcase with its sump ribbed for oil cooling. Drive to the vertical Cozette through bevel gears was a built-in feature, as were individual exhaust ports for optimum scavenging. At his Paris works at the Quai d'Asnières, from 1928 Raymond Jean Henri Siran powered his low-slung D'Yrsan sports cars with the Type K Ruby.

Siran also fitted several D'Yrsans with exotic engines from Strasbourg. There August Michel, after a flying career in the Great War, established Aviation Michel to trade, at first, in aero engines and then to manufacture them. His final offering, the AM16, was an in-line six for light aviation in alternative capacities of 998, 1188 and 1394cc. Their respective powers at 5000rpm were 75, 82 and 87bhp thanks to Cozette blowers mounted vertically at the front and supplying inlet ports along the right-hand side. Michel dealt with the resultant high revolutions – unsuitable for propellers – by fitting a 3.0 to one epicyclic reduction gear.

All-aluminium and impressively light, the largest weighing only 286 pounds, Michel's AM16 sixes had great potential as sports-car engines. They had twin overhead camshafts operating through rocker arms and a roller-bearing bottom end with a built-up crankshaft carried by five main bearings. Heads made in pairs were bolted to the block with its inserted liners. The units supplied to D'Yrsan were larger than Michel's aero engines at 1455 and 1646 cc, thanks to longer strokes, the biggest certainly surpassing 100bhp. But by 1930 D'Yrsan was no longer an active car maker.

At Boulogne-sur-Seine Louis Lefèvre produced La Perle cars and occasionally raced them. Introduced in 1924, his most ambitious model was powered by a 1.5-litre overhead-camshaft six. Although he stopped making La Perles in series in 1927, in 1930 Lefèvre updated his racing version of the six with the help of a Cozette. This raised its maximum speed from 85 to 105mph. Paris-based Sizaire Frères also fitted a Cozette, between its engine and the firewall, to a lightened sprint model. This blown version of Sizaire's two-litre four had double the usual quota of eight valves. No production ensued.

At Unieux in the Loire region proprietary engines were manufactured by Jacob Hollzer's Compagnie Industrielle des Moteurs à Explosion, or CIME, where Daniel Perrier was the young chief engineer. CIME engines powered some of the EHP cars produced by Henri Précloux, who began manufacture in 1921. Taken over in 1923 by the company that produced its bodies, EHP began operating from Courbevoie.

For a sporting model introduced in 1925, the DT Spécial, Précloux obtained a special 1.5-litre engine from CIME. Cozette-blown, as shown by their M1C designation, such engines powered only a handful of EHPs. In the hands of G W Olive one such competed at Brooklands, where it won the President's Gold Cup race with an average of 89.05mph and a best lap of 94.86mph. Olive was still racing his blue EHP there in 1933. CIME also supplied 1.1-litre blown engines, one of

Carburetters to Compressors: René Cozette's Contribution

5. This Cozette installation, with the blower high above the engine and driven by a vertical shaft, has so far escaped identification. But it is certainly interesting!

6. Starting to work with with the Cozette in 1927, Frazer Nash began fitting its cars with Cozette-blown Anzani engines from 1928

7. One of Ernest Eldridge's twin-cam racing engines of 1926 was Cozette-supercharged, the blower being vertically mounted. Power of 122bhp at 5600rpm was quoted

which powered an EHP to fifth at Le Mans in 1927. In that year the firm exhibited a Cozette-blown road car at Paris, this time with a SCAP engine.

Another customer for CIME's engines was Paris-based Alphi. In 1928 it supplied a 1.5-litre four for a sports Alphi, which competed but retired at Le Mans that year. More ambitious was a CIME 1.5-litre six for an open-wheeled Alphi which was supercharged to compete in Grands Prix. Its record of non-starting in such contests included the French GPs of 1929 and 1930, for Edouard Brisson and Jean Poniato respectively, and the 1930 Lyon Grand Prix, where it non-started for Jean Poniatowski. Alphi made two more cars before downing tools.

Aviation in the form of propeller production was the origin of the Hélices Ratier company in the Paris district of Montrouge, near the Porte d'Orléans. After the war Paulin-Jean-Pierre Ratier diversified into the production of telephones, toys, electrical appliances and, for a few years in the late 1920s, sports cars. Although its four-cylinder engine was ambitious, with shaft-driven overhead camshaft, finger cam followers, staggered valves for cross-flow porting, monobloc construction and a built-up roller-bearing crankshaft, the Ratier was of only 746cc. Naturally aspirated it produced 34bhp, considered good at the time.

In 1927 Ratier fitted a Number 6 Cozette compressor. Powered by the camshaft drive at the engine's front, the blower protruded horizontally from its right-hand side. A Cozette carburetter was above it, delivering direct to a manifold integral with the block. An extension from the supercharger's internal drive shaft turned a small oil pump dedicated to the blower, a standard Cozette feature drawing from a separate reservoir. Output on petrol was 46bhp, while the use of a mixture with benzol allowed 61bhp at 6000rpm. In this trim the Ratier set records in its international Class H, including five miles at 96.57mph.

A disadvantage of the compressor placement, said a Ratier owner, was that it was in the way of the front spark plugs and thus forced the use of a small plug size which was available in only a few heat ranges, making it difficult to find one that could cope with oiling-up when cold. 'The engine was a glutton for revs,' he related, 'but had practically no power low down. This, coupled with the fact that it was quite the noisiest car I have ever heard, in spite of two silencers, made driving rather a nightmare. Another source of annoyance was the Ratier's habit of catching fire on every conceivable occasion.' Having made some 30 Ratiers, Paulin-Jean-Pierre developed his 1925 patent for a variable-pitch propeller and became a major supplier of that to the French Armée de l'Air.

Easily the most viable and practicable of the early vane-type compressors, René Cozette's invention quickly found applications across the Channel. At Willesden in west London, with good links to rail and the Grand Union Canal, Cozette Services was established in Scrubs Lane to attend to the needs of British customers for both carburetters and superchargers.

Also in Scrubs Lane was the British Anzani Engine Company Ltd, set up by Alexandre Anzani – as Alessandro Anzani was known in his adopted French homeland – to produce his popular aero and car engines for the UK market. After a financial restructuring led by construction-company director Eric Burt, it was known as British Anzani Engineering.

Burt, then racing an Anzani-engined Aston Martin, suggested that the meritorious Cozette come under a British umbrella. In 1927 arrangements were made for a neighbour in Scrubs Lane, L T Delaney & Sons Ltd, to become the UK agency for Cozette products and designs. Active in the motor industry as an agent and engineer, Luke Terence Delaney was well placed to

1. Lea-Francis began experimenting with the Cozette in its *Lobster* racing car of 1926, finding it an excellent adjunct to better performance

2. Based on its work with the *Lobster*, Lea-Francis introduced its Hyper model in 1927 with its Meadows four fitted with a No 8 Cozette. No 9 was used for racing

serve the interests of all parties. Cozettes soon became familiar fittings on British road and racing cars.

No prisoner of his original concept, René Cozette applied for a patent on a new type of charger in February of 1927. Although still vane-type, it ingeniously fitted three vanes in slots through its eccentric rotor in such a way that they divided the delivery gases into six chambers. With this design, he said, 'the resulting centrifugal force is considerably diminished' – enough to allow elimination of his false rotor. Covered by the patent was the method of machining the 'cardioid' or 'heart-shaped' variation from a true circle that the geometry of this system described.

A potential breakthrough for Cozette's compressor was engineered by fellow Frenchman Albert Champion, leading to the trip to America mentioned earlier. After experimenting with a series-built supercharged Mercedes, General Motors decided that further investigation would be worthwhile. One of its component divisions, AC Spark Plug, began work in partnership with one of GM's car divisions, the Oakland Motor Car Company in Pontiac, Michigan, to explore forced induction's potential.

AC, which had been GM-owned since 1909, bore the initials of its founder, Albert Champion. Loyal to the nation of his nativity, Champion bought the US rights to the French Cozette compressor and brought its inventor to the New World to work on it with his own people. Champion had a highly qualified American on the project too, one he liked personally and had lured away from a government job in aviation supercharging at McCook Field to head his supercharger development. This was David Gregg, who had designed Fred Duesenberg's pioneering racing compressors.

In 1928 Gregg reported on AC's work in a

> 'It is no exaggeration to say that an important phase of supercharging in France died with René Cozette'

paper to the Indiana Section of the Society of Automotive Engineers. Although the charger used was clearly based on a Cozette design, Gregg did not identify AC's partner in his paper, calling the unit 'a supercharger developed in our laboratories'.

As a basis for the trials AC and Oakland chose the smallest six in their inventory, the three-litre sidevalve engine powering the Pontiac, a companion marque introduced in 1926. Nominated was its export version because it had magneto ignition powered by a chain drive. Adapting the chain to power the compressor gave flexibility in settling on a suitable drive ratio. 'This was not a quiet drive,' Gregg admitted, 'But it answered our requirements because we were more interested in securing data and results without great expense than in designing a drive such as might be used on a supercharged car in production.' He mounted the blower at the front of the engine's right side, high above the inlet ports 'to prevent loading under idling conditions'.

Working with René Cozette, AC experimentally varied the vane-type blower's porting so that it functioned both with and without internal compression. The experienced David Gregg tested various positions for the carburetter before and after the Cozette. Drawing from the carburetter, he said, 'The fuel mixture passing through the supercharger seemed to aid in distributing the oil to all the working parts and lowered the operating temperature of the supercharger, though the final temperature of the mixture was the same as that of the first system. Each system gave about the same power increase and each showed good accelerating characteristics.'

In fact the Pontiac engine's output, against an unblown 44bhp at 2800rpm, was 64bhp with the Cozette before the carburetter and 70bhp when it was fed by the carburetter, both at 3000rpm. 'The power increase varied,' said Gregg, 'from 35 per cent at 1000rpm to 59 per cent at 3000rpm.' The few changes to the base engine included removing the hot spot between inlet and exhaust manifolds after overheating of the inlet charge gave rise to preignition. Boost pressures of up to 0.46 bar required fuelling with a blend of 20 per cent benzol with 80 per cent ethyl petrol.

Turning at a step-up ratio of 1.28 to one, the compressor's power demand rose from

Carburetters to Compressors: René Cozette's Contribution

0.5bhp at idling speed to 11bhp at 3000rpm. 'We did not use a blow-off valve for preventing damage from backfiring,' Gregg told the SAE, 'And had no trouble on that account, although the blow-off valve is used on many supercharged foreign cars and apparently is found to be necessary. The chain drive was noisy, but the supercharger itself was quiet and had no objectionable note.

'The supercharged engine was installed in a two-door sedan,' David Gregg reported, 'And was taken to the General Motors Proving Ground and tested beside a similar model equipped with a high-compression engine using ethyl gasoline.' These trials showed acceleration times from 15 to 50mph of 21.0 seconds for the latter and 12.7 seconds for the Cozette version. Hill-climbing was dramatically enhanced by the supercharger. 'The gain in maximum speed was not so great as might be expected from first consideration,' the engineer reported, this being a margin of five miles per hour above the standard car's 64mph. Unlike many supercharger trials and applications, no attempt had been made to use the blower to extend the engine's range of revolutions beyond the standard car's 3000rpm.

'It is difficult to portray by mere figures the sensation of power and fleetness obtained while driving a supercharged car,' Gregg told his SAE audience. 'At road speeds from 35 to 50mph the car was a delight to drive, accelerating past other cars with ease. At all speeds up to the maximum there was no labouring of the engine; power always seemed to be in reserve. The car ran smoothly in high gear at 10mph. Starting and accelerating in second gear provided a getaway that left other cars far behind.'

Here were results from a first tentative trial that showed that a three-litre engine, small by American standards, could be made to give big-car performance. The results seemed to open the US market to AC Spark Plug's version of René Cozette's supercharger, replicating in the New World its considerable success in the Old. Fate seemed to play a hand, however. In 1927, at the age of 49, Albert Champion died of a heart attack in Paris. The loss of his pioneering spirit was a blow to the project. And, with America's ensuing slump into a depression, GM's executives had other matters on their minds.

In 1927 René Cozette decided to chance his hand with a car of his own design. Its engine was a two-stroke on the lines of the Czech CZ and Fiat's 451. Two opposing pistons in each cylinder were connected to crankshafts at top and bottom, geared together to drive the clutch. Bevel gears at the nose of the bottom crank turned Marelli magnetos at both left and right that fired spark plugs on both sides of the chambers, as well as driving a Cozette compressor mounted vertically – his preferred disposition. Its output supplied ports on the right side of the upper bores, while the lower pistons uncovered exhaust ports on both sides. As in other such two-strokes the blower both scavenged the cylinders and provided a modicum of boost.

With a 1.1-litre capacity this was a lively engine for the Voiturette category. The Cozette racer appeared in 1928, a slim single-seater with transverse-leaf springing. While Cozette himself raced it in several events without result, its greatest success came in the hands of Ghyka Cantacuzéne. A Paris-dwelling nobleman of the renowned Romanian Cantacuzino family, this wealthy sportsman took the Cozette to Montlhéry to

3. When Sunbeam looked for better performance from its three-litre engine in 1929 it fitted a Cozette, which brought a 53 per cent power improvement. It was alongside the engine like that of the Guyot

4. On the right is a supercharged BNC. With its 1100cc SCAP engine fitted with a No 7 Cozette, the little car could reach 85mph

attack international class records. Over three days he notched up new marks at speeds up to 103mph.

In 1929 René Cozette brought his eponymous car back to Montlhéry to tackle additional records. On the steeply banked concrete bowl he lost control, evidently the result of a steering failure, and crashed fatally. At only 34 years of age one of France's most creative engineering talents was lost at the height of his powers. The last of the 73 patents granted to Cozette, a novel form of change-speed gear, was awarded to him posthumously.

It is no exaggeration to say that an important phase of supercharging in France died with René Cozette. To be sure others in France, such as Albert Lory at Delage and – grudgingly – Ettore Bugatti, adopted and exploited forced induction. But Cozette's was an unique knack for the use of superchargers to enhance performance. If French builders of racing cars in the 1930s seemed to have lost the will to use boost as part of their competitive armoury – as they did – the absence from the scene of René Cozette was undoubtedly the reason.

For this special edition of The Automobile *Karl Ludvigsen has adapted this article from his work in progress,* The Supercharging Story, *to be published by Haynes in 2013.*

THE SCHOOL GARAGE
Martin J. Daly (Est 1979)

1961 AUSTIN HEALEY 3000 MK II FAST ROAD SPEC. Recently rebuilt to Concours standards. Once a road/race car with FIVA papers, now a fast road example, ideal for Tour-aut, etc. Stunning and as new. Please contact us for full spec sheet............**£65,000**

1935 ALVIS FIREBIRD, 2 SEAT SPORT TOURER. Stunning example, very useable, great history, and very rare......................**£64,950**

1952 ROVER P4 75. Masons Black with original red leather trim. Original, tool Kit, 4 owners, and 41000 miles only from new. Full documented history, original Log book. unrestored, time warp, original, example. Stunning...........................**£12950**.

1922 BENTLEY 3 LITRE RED LABEL BOB TAIL. Chassis number 91. (Reg. XL 3595.) Stunning and very original, with superb patination. Early racing history, known to BRDC since 1935. The best driving example we have had in years.......................................**£POA**

1969 JENSEN INTERCEPTOR 1. The purest of all Interceptors.Silver with original Black leather, series 3 wheels, and Original set of Rostyles.44000 miles only, superb history file,Stunning, rust free example..**£39,950**

1939 BSA SCOUT SPORTS TOURER. Last owner 12 years, restored to concours standards several years ago. Original Middlesex number, all books, tools, etc. The finest we have seen, and very rare. Not to be confused with a similar model.............**£24,950**

1954 SUNBEAM TALBOT 90 CONVERTIBLE Finished in white with original red leather trim & new black mohair hood. Last owner 20 years and in superb driving condition. Although not Concours, the bodywork has one or two cosmetic paint areas, the chrome work is perfect, the engine and gearbox are superb. For sale as it is or we can bring car to Concours standards should you require.........................**£16,950**

1914 DARRACQ V14. (Two seater with Dickey seat). French Blue with back leather trim. Superb Original, matching numbers example. In pefect order with stunning history........................**£64,950**

1949 MG TC. Black with red leather and tan mohair hood. Aero screens, 100 Point Concours Restoration, and the best you will find........................**£44,950**

1908 SEARS MODEL J HORSELESS CARRIAGE. Twin cylinder with twin exhaust. In superb, original, condition and perfect working order. First owner till 1952, has been in 2 museums since. Very rare with superb history and period photographs from its early days in Atlanta....................**£49,950**

1935 AUSTIN 7 MILITARY TOURER. Very very rare, Chassis 228208, in superb original order. Only two are known to exist in working order, this is the best. Superb history file and period photos, it is a collector's dream. It has new, custom made, weather equipment and drives perfectly. *NOT* to be confused with normal Austin 7 examples:..**£16,950**

We are the new Official Dealers for the North-West and Midlands for the 'New' **PROTEUS C-TYPE-JAGUAR.** The best C-Type, we think, ever built to date (after the original). Full alloy body, fuel injected Jaguar 4.2 Ltr engine. Plus, plus, plus NOT a kit, but a genuine hand built car, with modern brakes, suspension, gear box, diff, with the finest leather trim, built to the highest of standards. **From £89,950**
Please call us today or visit our website.

1967 JAGUAR 3.8 MK2. Manual/Overdrive. Original matching numbers UK example Opalesant Blue with blue leather, restored to concours winning standards (and one previous owner). 41,700 miles only from new, and could well be the best in the UK..**£44,950**

1954 MERCEDES 300 W 186 (Adenauer) Very rare manual floor-change example, prepped for classic rallying (Gullwing spec engine with twin Solex carbs. Stainless steel exhaust, stainless steel fuel tank, sump guard and twin spot lamps. Fiva papers past and present available. This 3 owner motor car (first owner 38 years) is finished in its original black with grey leather trim and, although not Concours, is in remarkable original condition.......................**£69,950**

1954 MG TF MATCHING NUMBERS RHD UK Car. vory white with cherry red leather and black Mohair weather equipment. Restored to Concours-winning-condition, by its last owner of over 30 years. Original but Uprated Engine, and Also Unleaded. The Finest to be seen on the open market in years. Truely Stunning...**£34,950**

1957 ROLLS ROYCE SILVER CLOUD 1. Two tone sage green with beige leather, PAS, original RHD example, with all books, tools etc, known by us for many years, and sold by us to its last collector owner. 72000 Miles only, original build sheets in history file, lovely. Original, Stunning, example..............**£36,950**

1956 ROLLS ROYCE SILVER CLOUD 1. Oxford blue over Ascot grey with grey leather. A stunning 2 owner car with 18,000 genuine miles from new. Stored for c30 years and recently restored to a very high standard. Very seldom do you come across examples like this. All books, tools etc, original buff logbook. **£49,950**

PRESTIGE & CLASSIC CAR SPECIALISTS
(GENEROUS CASH DISCOUNT WITH NO TRADE-IN)

Tel.01663 733209.   **Mobile.07767 617507.**
47 Buxton Road, Whaley Bridge, High Peak, Derbyshire SK23 7HX.   www.classiccarshop.co.uk

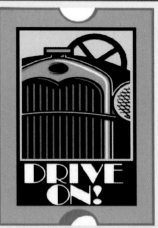

Gavin McGuire's Fine Automobiles
www.gavinmcguire.co.uk

These are the kinds of cars I sell
Please visit my web site for more information:
www.gavinmcguire.co.uk

Commission sales!
The best option for buyers and sellers
Phone/fax: **01892 770310**
Mobile: **07770 316482**
E-mail: gavin.mcguire@virgin.net
In south East England near M25, junction 6.
Close to London Gatwick

1901 BOYER 6.5 HP SINGLE - £PLEASE ASK

1903 RAMBLER 8 HP SINGLE - £PLEASE ASK

1903 RICHARD-BRASIER 12HP TWIN. - £P. ASK

1904 CADILLAC MODEL B - £PLEASE ASK

1904 DARRACQ 15 HP 4 CYLINDER - £PLEASE ASK

1904 DE DION 10HP TWIN CYL. - £PLEASE ASK

1905 SUNBEAM 12/14 - £PLEASE ASK

1907 MINERVA TYPE M 24 HP - £P. ASK

1913 WOLSELEY 24/30 HP - £PLEASE ASK

1913 AUSTRO-DAIMLER - £PLEASE ASK

1915 NAPIER 20 HP - £PLEASE ASK

1920 TALBOT 25/50 - £PLEASE ASK

1934 LAGONDA 3 LITRE - £PLEASE ASK

1934 LAGONDA M45 - £PLEASE ASK

1934 BENTLEY 3½ / 4¼ LITRE - £P. ASK

The Phantom of Love

You can blame Woolworth's director John Ben 'Surefire' Snow and Marie Antoinette equally for the baroque extravagance that is the interior of Rolls-Royce Phantom 76TC, commissioned in 1926 by C W Gasque as a surprise present for his wife Maude.
David Burgess-Wise tells the story

Dynamic New Yorker Surefire Snow had been personally asked by Frank W Woolworth, founder of the eponymous 'Five and Dime' stores, to help establish the American company's British chain of 'Threepenny and Sixpenny' Woolworth's shops. Frank's cousin Fred Woolworth had opened the firm's first British store in Liverpool in 1909. Snow, an astute buyer, arrived in Britain in 1910, six months after that first shop had opened, and set about expanding the company network.

Working closely with Woolworth's only English director, William Stephenson, Surefire Snow – appointed to the board in 1910 – helped build the British Woolworth's chain to 28 stores by September, 1913, and was appointed buyer for the Group, with an office at the company's rented London offices in Oxford Street.

In the autumn of 1913, Snow was invited to London and offered the prestigious role of Buyer, responsible for china, glassware, celluloid toys and novelties. When war broke out in August, 1914, he had quickly to find new suppliers for items previously purchased from Continental sources. He did so well that in 1915, despite the hostilities, British profits exceeded £4 million, helped to no small extent by Snow's 'Sixpenny Pops' range of popular song books.

Made wealthy by dividends on his shareholding in Woolworth's and an annual pay package of £50,000 (the equivalent of some £3.5 million today), Surefire Snow moved into a luxury flat in Clarendon Court, Maida Vale, and made it his London base.
A keen polo player, he bought the Highfield stud farm in Hertfordshire and developed a racing stable. At weekends he entertained company executives, expatriate Americans and members of high society at his farm. And he learnt to drive, and bought himself the first of a series of Rolls-Royce motor cars, which he used to visit his stores and suppliers.

A mutual friend introduced him to J H Barnett, owner of the old-established Wolverhampton coachbuilders Charles Clark & Sons, founded as coach and carriage builders in 1839 with impressive two-storey premises in Chapel Ash, a main thoroughfare on the west side of the city centre. Early in the 20th century, the company had realised that the future lay with the horseless carriage and moved into motor car sales and bodybuilding. After the owner, Charles Clark, died in January, 1911, Barnett took over the business. He carried on coachbuilding – concentrating on high-class work, principally on Rolls-Royce chassis, though

a handsome Sunbeam vee-fronted all-weather Clark coupé was illustrated in *The Autocar* in 1921 – while building up the retail side of the business.

Barnett's credentials were impressive. In 1902 he had been Herbert Austin's first pupil at the old Wolseley Works in Adderley Park, Birmingham, and was, courageously, the first man to join him when Austin set up his own business in 1905 on a shoestring budget of £15,000 after falling out with the board of the controlling Vickers company.

"Although I left Austin some years later," recalled Barnett in 1958, "Our friendship lasted until his death." And from Austin, said Barnett: "I acquired a love of craftsmanship and the ideal that nothing was good enough but the best. He was a man who did not consider cost where the question of the highest quality workmanship was concerned."

That philosophy obviously chimed with Surefire Snow, who had a number of bespoke bodies built by Clark & Son. Typical was the Silver Ghost he ordered in 1920, which was described by *The Autocar* as 'original… yet it is by no means blatant'.

Technically, Snow's Ghost – naturally, it was painted white – was a 'V-fronted, curved-back, interior-drive saloon of well-balanced appearance… with an harmoniously curved scuttle,' but in terms of finish and trim it was quite extraordinary. Beneath the scuttle the panelling was frosted aluminium, as was the vee-shaped instrument board. The roof and the upper part of the body were panelled in sycamore, with the roof finished in a marquetry star pattern. The upholstery was a reproduction of a Charles I tapestry and the frameless windows were hung with cream silk curtains suspended on silver rods. Moreover, added Barnett, "This man's stipulation was that, in addition to special upholstery, all fittings such as door handles, speedometer and instrument cases, etc, should be of hallmarked silver, and Elkington & Co made me

1. The Gasque Phantom has never been restored, but the exterior underwent one or two alterations in Stanley Sears's ownership

The Phantom of Love

2. The relatively modest exterior gives no hint, except via Sears's upgrades, to the splendours within. The car has covered less than 9000 miles in the 86 years since it was built, averaging just over 100 per annum

several sets to my designs."

Snow's exotic Silver Ghosts must have made a deep impression on his fellow Woolworth director Clarence Warren Gasque, another expatriate American, who Surefire introduced to Barnett. In the summer of 1926, Gasque commissioned Clark & Son to build the car that would become known as the Phantom of Love.

Recalled Barnett: "As I believe is often the case with Americans, this gentleman wanted a car for his wife which must be different to anything else and also better. He would not stipulate what he wanted except that the design must be French and left everything to me including price. Like the former customer, he absolutely refused to come and see the car while it was being built, and neither of them saw the cars until I delivered them in London. I have thought since that there was a great risk in this, but I am glad to say that in no case did I ever have any complaint either as to workmanship or price, except that one car came back because he felt I could have put in more silver fittings, but this addition he paid without demur."

Gasque's rise to the top of the Woolworth organisation had been atypically sudden – he had, says Paul Seaton of the Woolworths Museum, "mysteriously appeared in 1919 without serving in-store and without any explanation in the Company's Minute Books" – but is probably explained by the fact that he was the brother-in-law of Hubert Templeton Parson, appointed president of Woolworth in 1919 after the unexpected death of company founder Frank Winfield Woolworth. Parson's role was essentially to keep the store managers happy and turning in good results; when it came to major decisions, board members expected the founder's brother, chairman Charles Sumner Woolworth, to make them.

Parson and his wife Maysie were addicted to the high life, trying to emulate the gilded lifestyle of the company's founders, their ultimate extravagance a

Automobile 127

huge, 300ft-wide mansion called Shadow Lawn in West Long Branch, New Jersey, designed by fashionable Philadelphia architects Horace Trumbauer and Julian Abele, with 130 rooms elaborately decorated in American Beaux Arts style. It was built in 1929 at a cost of $10.5 million after the previous wood-frame house on the site – a former summer residence of President Woodrow Wilson with a mere 52 rooms – had burnt down. Unfortunately, Parson had just expended $1 million on its refurbishment.

The year he became president of Woolworth's, Parson (whose luxurious tastes were ridiculed by his colleagues) sent brother-in-law and new board member Gasque, an accountant like himself, to London as director and secretary of the F W Woolworth & Co Ltd 'Threepenny and Sixpenny' Stores. Gasque's mission was "to count the pennies and keep a close eye on his wayward London subsidiary," says Paul Seaton, adding "By all accounts Gasque was a nice chap who did a good job but kept a low profile, and was left to get on with things by his fellow board members… Executives in London treated him as a spy in the camp and left the bean-counter in his office sending secret cables back to New York. They didn't worry too much about what he had to say, as it was well known that Parson was isolated as President… Like all of the British directors, with the firm doing phenomenally well, Gasque quickly built up share options worth millions at today's values."

Clarence Gasque shared his brother-in-law's taste for opulence. His vast mid-Victorian house, The Elms, in Spaniards Road on Hampstead Heath, the former home of art dealer Sir Joseph Duveen, was built in a bastard 17th century style with Dutch gables and a prominent corner turret; it stood, 'surrounded by green lawns and beautiful old trees', in sweeping acres entered through elaborate wrought iron gates.

It's evident that Gasque wanted the Rolls-Royce he had commissioned to outshine anything that Surefire Snow had bespoken, and at first coachbuilder Barnett was at a loss to know where to start. He travelled up to London seeking inspiration at the Victoria & Albert Museum in South Kensington, where he saw a 'very delightful little Sedan Chair which had once belonged to Marie Antoinette and which had a painted ceiling'.

Given the unfortunate Queen Marie Antoinette's notoriety for extravagant living, it was perhaps a happy omen, and Barnett began drawing up plans for a Coupé de Ville whose sober exterior concealed perhaps the most exotic interior ever seen in a motor car, among its unusual features a painted ceiling decorated with naked amoretti frolicking among swags of roses. The carved and gilded cornice that edged it concealed lights that illuminated the passenger compartment. Further lighting was supplied by gilded cherubs holding flambeaux in amber glass at either rear corner.

At Wolverhampton, recalled Clark in 1958, "I had a very small staff, but a very good working foreman, and I made what sketches and plans he required… All the interior woodwork was done there, but some of the carving was done in London. The panels and cabinets we made entirely, but the painting was done by a Frenchman in London of whom I have lost trace.

> "It's obvious that Gasque wanted the Rolls-Royce he had commissioned to outshine anything that Surefire Snow had bespoken"

The interior metal fittings were made by Elkingtons to our design, and I think I still have some of the original castings."

With panelling in highly-polished quartered satinwood veneers decorated with painted arabesques and oval medallions, the passenger compartment was more like a Versailles throne room than the interior of a motor car, the effect enhanced by the rear seat, which was upholstered in the finest tapestry embroidered with scenes in the manner of Rococo French painter Jean-Honoré Fragonard, as were the fold-down inward-facing occasional seats, which were concealed in cupboards either side of a bow-fronted drinks cabinet mounted on the division, which gave the effect of an antique Louis XV commode. Said Barnett: "The tapestry was made by Aubusson, and I well remember that it was a very hazardous job to make patterns for this before the job was really started, but as it took over nine months to make, we had to get it in hand at an early date. This tapestry cost me over £500." Which in 1926 was the price of a new 12/50 Alvis…

When Gasque requested that Barnett add a crest to the car's rear doors, the Wulfrunian coachbuilder politely remarked that Americans didn't have coats of arms. However, Gasque responded that his family was of French extraction, so Barnett devised a pseudo-heraldic crest that "pleased everyone"…

The finished car, which had cost some £4500 in total, was featured in a number of magazines, among them *Motor Body Building and Vehicle Construction* for May, 1927, which enthused: 'We have on more than one occasion referred to the artistic workmanship and refined taste in decoration for an interior period treatment displayed in coachwork designed by Mr J H Barnett, and built under his direction in the works of Messrs Chas Clark & Sons, Ltd, Chapel Ash, Wolverhampton. The Coupé de Ville mounted on a New Phantom Rolls-Royce chassis is one of their latest productions built to the order of a lady that shows unmistakable evidence that there is a demand for a distinctive class of coachwork that calls for a high standard of craftsmanship in design, construction and finish…

'At first sight it may appear as if ornamentation of an elaborate character had been the object rather than comfort and utility, but this is not the fact, as a reference to the illustration will at once convince our readers that there has been great desire to combine comfort and neatness with elegance…

'In the production of this fine specimen of motor bodywork Messrs Chas Clark & Sons have accomplished an achievement in combining a perfection of fine work with utility, which will enhance their reputation amongst coachbuilders.'

But soon after he had delivered Gasque's car, Barnett realised that while the work had been intensely interesting, it had shown little profit. And since the demand for such coachwork was waning, he closed the firm's bodybuilding department and concentrated on car sales. Around 1954, Barnett sold Charles Clark & Son to the group headed by Sir Alfred Owen. The company survived as a Rover agency until the early 1990s; the Charles Clark name has since been revived

1. The Phantom's painted ceiling was inspired by a sedan chair which had once belonged to Marie Antoinette

2. As illustrated in a contemporary coachbuilding magazine, Mrs Clarence Gasque's conveyance looked inside like something straight out of Versailles

3. Gasque and his wife Maude pose beside their new possession, considered at the time to be one of the most magnificently fitted-out luxury cars ever built

The Phantom of Love

as a Toyota dealership.

Sadly, Clarence Gasque died at the comparatively young age of 54 on 12th October, 1928, only 18 months after presenting this costly love token to his Maude. While his widow could have lived on in idle luxury on his millions, instead she plunged headlong into the ascetic world of vegetarianism and animal welfare, eventually becoming president of and major donor to the International Vegetarian Union, vice-president of the Vegetarian Society and an advocate of the offbeat teachings of 'master mind' Otoman Zar-Adusht Ha'nish, Czar of Adusht (born Otto Zachariah Hanisch), and his 'syncretistic religious health movement' Mazdaznan, in which colonic irrigation and breathing exercises played a major role.

While she travelled the world in her mission to promote her 'overriding desire to better mankind and relieve the creature kingdom of its suffering', Maude Gasque put her Phantom of Love into storage in 1937, and Barnett sadly noted: "The last time the car came to us for repainting, the interior fittings and the veneered panels were getting into rather a bad state due to the fact that the car had been kept in an unheated garage, and this did not seem to concern them when I pointed it out to them."

Post-war, Maude Gasque motored in a 1950 Mk V Jaguar drophead with spatted rear wheels, a rare model of which only 28 were built. She enthusiastically preached her eccentric doctrine at meetings of the International Vegetarian Union, where she was known as Mother Gloria, speaking under a huge banner proclaiming 'Vegetarians of all nations unite, you have a world to gain for justice, kindness, health, happiness, peace, progress and prosperity'...

She died on 23rd December, 1959. At that time, she owned homes in Tecate (on the Mexico-USA border) and in Los Angeles, California; her former Hampstead residence, The Elms, had become a hospice in 1957.

In 1952, the Phantom, with less than 10,000 miles on its odometer, was bought by that great Rolls-Royce collector Stanley Sears. Curiously, he wrote that 'Unfortunately, I did not know of the car's existence until it got into the hands of a London dealer, who bought it direct from the executors at a very low price when Mr Gasque died. I am afraid I had to pay through the nose for it.'

Now the nub of that statement is demonstrably untrue, for Clarence Gasque had been dead nearly 24 years, and Maude was still very much alive in 1952. Nor does it match the tale spun by the London dealer, Metcalfe & Mundy of Adam & Eve Mews, Kensington, whose advertisement – which only seems to have appeared once – I found by trawling through the small ads in the 1952 issues of *The Autocar*; they claimed that 'the most fantastic car of all time' had been used 'by a nobleman for official functions only'...

I'm sure it's true that Stanley Sears did indeed pay through the nose to acquire the Phantom: but that didn't stop him from spending even more on putting his own stamp on Mr Barnett's masterpiece. Believing that the restrained exterior of the car failed to match up to the opulence of its interior, he had its rear quarters finished in simulated canework, painted the wheels straw colour and lined out the bonnet.

I recall first seeing the car in 1963 at Sears's home, Old Highlands in Bolney, Sussex, when I rode down in the rear of a friend's fabric-bodied Rolls-Royce 20 to attend the firing-up of his recently-acquired 1905 30hp

1. This Rococo clock, supported by a naked *putto*, graced the satinwood-veneered interior partition

2. Typically English in its proportions, the relatively restrained Coupé de Ville coachwork gave little hint of the treasures within

3. The drinks cabinet, bow fronted and veneered in satinwood, retains all its original fittings

4. The Aubusson tapestry works wove the door panels and seat covers in Louis XV style in the manner of Jean-Honoré Fragonard

5. The driver's compartment, as in most formal carriages of this type, was functional rather than flippant

six, then in chassis form, for a group of Rolls-Royce Owners' Club members.

Though he moved to Spain and sold the bulk of his collection in 1983, Stanley Sears kept the Gasque Phantom until 1986, when he sold it to a Japanese collector named Takihana. It passed through successive ownerships in Japan until December, 2001, when Akira Takei sold it to American dealer Edward Fallon of Cave Creek Classics in Phoenix, Arizona, who happened to be in Japan on business.

Fallon lost little time in selling the car on to prominent Pennsylvania collector Jack Rich. For the next few years, it became a regular on the concours circuit in America, winning many awards, including the Lucius Beebe Trophy at Pebble Beach in 2002. But the car began to suffer from exposure to the weather on the close-cropped lawns of the American concours circuit, and Rich sold it to English dealer Charles Howard, who had sought to own the car for over 30 years. Howard renovated the Phantom and fitted it with black wheel discs. From him the car went to Rolls-Royce specialists P & A Wood in Great Easton, Essex, who exhibited it at Rétromobile in Paris in February, 2004. It subsequently found a new owner who, after seven years of enjoying this remarkable machine, has decided to put it back on the market. So there it stands in the Great Easton showroom with less than 9000 miles on its silver-rimmed odometer. If there's a single car that really should be in the heritage collection of Rolls-Royce at Goodwood, it's this one – surely the ultimate example of the bespoke motor car.

Thanks to P & A Wood and to Paul Seaton of the Woolworths Museum

Symphony
A very special MG TA

Park Ward were responsible for many a striking body on Rolls-Royce and Bentley chassis. Did they only clothe large cars? Of course not. Among the big ones was one tiny MG. Charles Ward wanted a special body for his own car, so he made one himself. **Rutger Booy** tells the story

William Park and Charles Ward were both experienced craftsmen. They had worked at F W Berwick & Sons, at the Sizaire-Berwick motor car factory in Park Royal, north-west London. In 1919 they decided to leave their jobs and go into business together as coachbuilders. Their enterprise, known as Park Ward & Co, was set up in Willesden, also in north-west London, which was in due course to become a centre of the coachbuilding trade.

Their business flourished. In 1920 they made their first body on a Rolls-Royce chassis and in 1923 the company started a long-lasting association with Bentley, for whom they bodied some 140 chassis during the 1920s. Park Ward also created many car bodies for Rolls-Royce. Of the 2940 20hp models that were built from 1922 until 1929, the company was responsible for bodying 325 of them, mostly standard sports saloons and limousine designs.

By the late 1920s Park Ward started to feel the effects of the recession. Output fell from 150 bodies to just 80 in 1931. Surely it must have come as a relief that, soon after Rolls-Royce had taken over Bentley, the Willesden company was asked to become the preferred supplier of Bentley saloon bodies. In 1933 Rolls-Royce even took a financial interest in the coachbuilder, but then began to express concerns about the quality of Park Ward's coachwork. This was problematic to say the least, with doors flying open at unwanted moments. So Park Ward started experimenting with an all-steel body. After a year of testing, the new solution was introduced in 1936. It turned out to be a success and was used on many Bentley chassis.

Still the main strength of Park Ward was the development of bespoke bodies. Peter J Wharton, who joined in 1934 and became chief draughtsman, later described their clients as: 'Gentlemen of impeccable taste, who ordered custom-made coachwork, were invariably conservative in outlook and liked the more traditional approach to coachbuilding.'

By that time Charles Ward's son, also called Charles, was working in the company. By his own admission Charles would have preferred to become a physician but, presumably under pressure from his father, he joined Park Ward and learnt the trade. Watching the many attractively-bodied automobiles leaving the works destined for wealthy clients made him yearn to have such a car for himself. However, being 30 years old and married, he found a Rolls-Royce or Bentley out of the question – they were much too expensive.

Charles settled on an MG TA as the ideal basis on which to design a body. It seemed a good choice. Introduced in June, 1936, the TA was the successor to the P-type, fitted with a bigger chassis and longer wheelbase to make it roomier and more comfortable than previous Midgets. Already MG had fallen victim to the Nuffield policy of harmonisation, and had to use as many standard Nuffield components as possible. This meant no more overhead camshaft engines were to be fitted; instead, a humbler overhead valve unit from the Morris group was used, in this case a Morris 10 with a capacity of 1292cc.

It is not known when Charles bought his TA, nor if it was acquired new or second-hand. Anyway, it was an affordable and easy to maintain automobile. One can

A discreet MG badge covering the aperture for the starting handle saves owner Hemmo de Groot answering repetitive questions

Symphony – a very special MG TA

Looking just like a scaled-down Bentley drophead of the immediate post-war era, *Symphony*'s revised styling betrays its constant upgrading while in Charles Ward's hands

only suppose he must already have had a good look at the MG TA then in production by his rival Salmons & Sons. In 1936 this coachbuilder from Newport Pagnell had signed a contract with MG in Abingdon to fit their elegant Tickford drophead coupé bodies on several MG models. One can safely assume Charles knew of this and thought he could do an even better job.

Although Charles Ward himself already had the skills to build a new body, he did have some help from others at the company. For instance, the aforementioned Peter Wharton, who worked in the design department, produced the rendering seen here. It's pretty similar to the finished car. Small differences can be noted in the length of the bonnet, the rounded shape of the doors, the longer luggage compartment and the discs on the wire wheels which were not on the actual car.

Being much smaller and altogether less substantial than one from a Rolls-Royce or Bentley, the MG's chassis must have presented many problems. All the lines used on the bigger bodies had to be scaled down. Although the chassis of the MG TA was strong enough for its original purpose, to carry the weight of the Park Ward coachwork he had in mind Charles had to strengthen it in several places. And to propel the new body the TA's slightly underpowered engine had to be tuned, so a low pressure Arnott supercharger was fitted.

All in all it took 11 months to build this one-off dream car, and during the construction its proud proprietor kept a photographic record of his work – not an everyday procedure back then. When the MG was finished Charles was understandably pleased with it. On the final page of his records he wrote: 'An idea, a year's hard work and a life's ambition made.'

Now let's fast forward to the 1970s, when Dutchman Hemmo de Groot enters the story. Like many a young man then and now, he wanted a real sports car. After considering several makes Hemmo realized he was very much Anglo-minded, so he bought himself a second-hand MGB. Having joined the MG Car Club in 1974 Hemmo became even more enthusiastic and restored the MGB to concours condition. Over the years, looking at other members' cars, he became interested in earlier MGs and decided he wanted one. Not a true Vintage MG, but one that could be used in daily traffic. So he started looking for a T-Type, preferably a prewar TB.

In 1985 a friend told him about an old, run-down MG that he knew lay mouldering in a shed somewhere in England. According to this friend, it was known as an MG Ward. Not knowing what to expect, Hemmo followed up the tip and thus found *Symphony*, although at that time he didn't know the car by that name. Whilst it looked like a total wreck, the aluminium coachwork seemed reasonable apart from a few dents and scratches. The woodwork and the chassis, too, looked well preserved, but the interior was in a very sorry state and the engine was lying in bits somewhere else in the shed. With the car came some photographs and papers, the pictures having been taken just after the war in front of the Park Ward buildings. From these Hemmo learned that Charles Ward was never finished with the MG and after the war tried out different designs: he made longer, more flowing wings; he altered the waistline by changing the chrome side strips; he transformed the hood from a cabriolet top into a drophead coupé and fitted the streamlined nose that can be seen in Peter Wharton's full-colour drawing. The

1. *Symphony* in her original form, with short wings and MG T-Type grille

2. The second version of *Symphony*, now with longer wings. It has the drophead instead of the earlier hood that disappeared below the body line

3. The Arnott supercharger

4. Under construction: a photograph from Charles Ward's personal collection

5. As found in a barn in 1985

MG became a real test-bed for new ideas. Look at a 1950s Bentley made by Park Ward and the similarity is obvious.

At one point, presumably in the late 1950s or early 1960s, the MG lost its glamour and was sold on. Eventually she ended up in this shed where she further deteriorated, getting ripe for the crusher. Yet, when Hemmo saw the flowing lines of the Park Ward bodywork and realised what a beauty she had once been, he knew on the spot this was the MG he wanted.

What followed was a body-off restoration that took the best part of seven years; Hemmo is a perfectionist and wanted his TA to be exactly right. But first he had to decide what form the car had to take: restore it with the chrome T-Type radiator shell or the painted streamlined nose? Being an MG enthusiast, Hemmo opted for the square front MG grille as this was Charles Ward's original design. He soon found that the grille that came with the car was not a standard one from a T-Type, but one that had been made wider to fit between the wings of the Park Ward body. Restoring this radiator shell was an endless hell of a job: it had to be chromed, repaired and re-chromed five times to get the correct, spotless finish.

Dismantling the car was done the proper way. Large assemblies that had been taken off the frame were stored in boxes; small parts in more than 150 plastic bags, each containing a description of what it was and its place on the car. Building up the chassis was the easy part. This was done in Hemmo's living room, so he could watch TV at the same time. The engine also got the full treatment. All moving parts were hardened, balanced and polished. When ready it was spray-painted in red lacquer. Restoring the coachwork proved not so easy. That could only be done with the help of a good friend in his garage. All in all, the restoration took almost 4000 hours.

When finally finished in 1992, the Park Ward MG TA looked a very pretty picture in two-tone dark and light blue. It had a friendly, sporting image, but something in this picture was missing and Hemmo could not put his finger on it. A year later, after a magazine had published an article about his restored MG, Hemmo was approached by Joy Ward, widow of Charles. She had heard of the restoration through an ex-colleague of her late husband's, who had commuted with him in the MG to and from the factory for almost six years. A visit to Joy Ward was soon arranged. She was a small, chic lady, then more than 80 years old, and she was delighted to see the MG again. She told Hemmo her husband had a name for it. He called her *Symphony*. This was the first time Hemmo had heard this name and from then on the TA was again known as *Symphony*. Joy Ward endlessly told Hemmo about the many happy trips she and her husband had made in the car together. When Hemmo opened the bonnet, she pointed at the engine and exclaimed: "I put the red paint on that!" She still had the photographic report that Charles had made while building *Symphony*. Photocopies were made and this allowed Hemmo to see what he had done wrong (but also what he did right) during the restoration.

Symphony – a very special MG TA

6. Drawing by Peter Wharton of the car's final post-war configuration (*Wiard Krook Collection*)

7. The MG's long bonnet is well filled by its engine, supercharger and accessories

8. Believe it or not, the little car in its revised form had push-button electric windows. Are these the first in a British car?

9. Unique: Park Ward's signature on the only MG they ever bodied

By now he had finally realised what had been troubling him for a long time. It was the square front of the MG that in his view didn't really harmonise with the flowing lines of *Symphony*'s rear end. The streamlined nose that the car had received just after the war matched perfectly with the flowing lines of the rest of the coachwork. Still, it hadn't been a bad decision to use the MG grille as by that time *Symphony* had won several awards in the Concours d'Elegance at Paleis het Loo in Holland and the one in Schwetzingen, Germany. But what to do? You don't just take a seven-year restoration apart again.

Hemmo's chance came when a few years later, in 2006, the chrome T-Type grille showed signs of fatigue and had to be replaced. Hemmo decided to rebuild the streamlined nose as used in 1946 and, with the help of a local specialist, started to create a new one: more easily said than done. What looked like a simple nose job meant the construction from scratch of a new grille, including a chrome surround to keep the gauze in place. The radiator needed to be moved, and there's another detail that shows what kind of perfectionist Hemmo is. Looking at the old photographs he noticed that the louvres in *Symphony*'s bonnet were now different. The ones on his version were straight and pointing to the outside; on the old, original pictures they were tilted and pointing to the inside. As a consequence a whole new bonnet had to be made – again, quite a job, but when it was finished Hemmo was satisfied that *Symphony* looked just like Charles Ward had wanted her to look. With one exception: Ward had made his own Park Ward badges, but had not thought to put an MG one back on. This bothered Hemmo. People were always asking what kind of car *Symphony* was, so there is now an MG badge covering the hole for the starting handle.

En route to a location for a photoshoot, Hemmo steered *Symphony* with ease through the heavy commuter traffic of his home town. The engine picked up responsively, the exhaust giving a deep, pleasant growl. Shifting through the gears was quite easy. According to Hemmo it's no problem to drive long distances in *Symphony*, as he had done, for instance, to the last Goodwood Revival or to Angoulême to watch the historic races. Truly a restoration well done.

The story doesn't end here, because Hemmo has clear thoughts about the future of *Symphony*. Once, at an MG meeting at the Silverstone circuit where he hadn't entered the Concours d'Elegance, someone had come up to him and said: "With a car like that you really must enter the Concours and let other people enjoy her, too!" These words set Hemmo thinking and made him realise that he's not only the owner of *Symphony* but, as with any beautiful work of art, he is also its caretaker for the future. This thought he puts into practice by being certain that any job that has to be done to *Symphony* is done well, thus making sure it has not to be repeated in the foreseeable future. He also collects all kinds of parts – for instance, adding recently the set of 16in wire wheels with which *Symphony* can be seen in several old photographs. These give the car a wider, sportier look, although normally Hemmo uses the smaller wheels with the rear spats in place.

Today the MG is more than 75 years old, but given Hemmo's view to the future it seems very likely that *Symphony*'s wonderful exhaust will be heard for at least another 75 years and, who knows, maybe many more.

At speed behind the wheel of his seven-year restoration project, Hemmo's face reflects his satisfaction

Symphony – a very special MG TA

1. The door panels were made of 10 separate elements, separated by contrasting piping coming out of one corner

2. The long, flowing wings give the MG an elegance rarely found in a small car

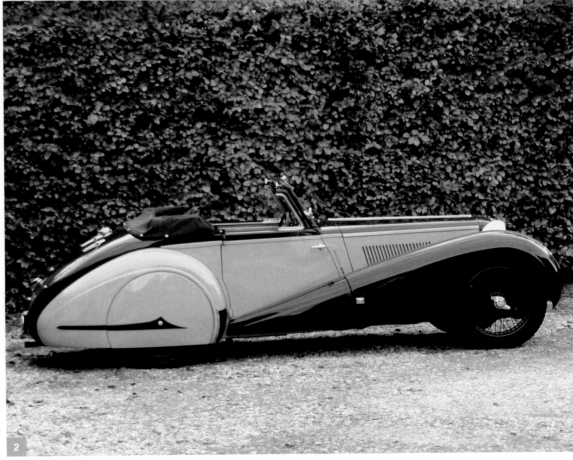

1947 Bentley MKVI Convertible by Park Ward.
Finished in Regal Red with Champagne hide interior, rare and elegant short winged version with encased spare wheel in offside wing. Masses of restoration history.
Please call for further details

1935 Sunbeam 6 Cyl 3.4 Shooting Brake By Mulliners
Serveral concourse first class wins to its credit.
Please call for more details

Bentley Turbo R Coachbuilt Double-Deck Hearse
'Highway to heaven' transportation for the man who thought he had all incidents covered, has never been used, for further details please call

1962 Bentley S2 Continental Convertible
Finished in Windsor Blue with matching hood, known history from new, has just returned from lengthy European tour. For further information please call

Laughton Investments

Lutterworth, Leicestershire

1932 Rolls-Royce, Phantom II Sedanca de Ville by Windovers
Same family ownership for 79 years. Finished in Dove Grey and Mediterranean Blue with Champagne hide interior. Please call for information pack

1933 Rolls-Royce 20/25 Gurney Nutting style 3-position Drop Head Coupe
£70,000 spent already, requires finishing for a result similar to the above.
Please call to discuss and for more detailed current photographs. £24,950

1937 Packard Super 8 Limousine
Superb example having been previously owned by a Scottish aristocratic family, information pack and more detailed photographs available. Please call

1925 Bentley 3 4.5 litre Speed Model Red Label Tourer
Fitted with overdrive, hydraulic brakes and ready for all events.
Priced competitively at £269,000

T: + 44 (0)116 240 2115   F: + 44 (0)116 240 4444   M: + 44 (0)7967 649761   Email: laughtoninvestments@gmail.com

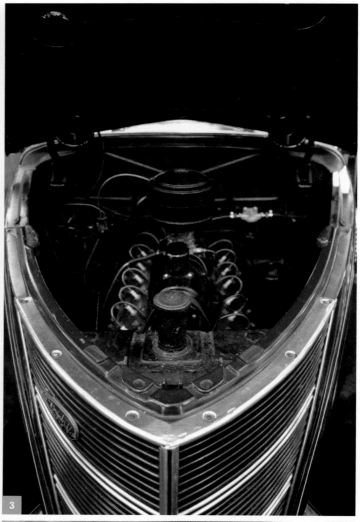

1. The interior of *The Automobile*'s 1937 Lincoln-Zephyr is entirely original. The dashboard retains its original semi-metallic painted, the shade being called Mercury. The waterfall-style centre section of the dashboard is illuminated at night, as well as the instruments

2. Stylish touches abound on the Lincoln-Zephyr, even down to the typography on the hubcaps

3. The 4.4-litre 75deg V-12 develops 110hp and can propel the car to 90mph in favourable conditions. This particular car, which has covered less than 100,000 miles from new, has been cosseted for its entire life and has never been subjected to a full restoration

4. The famous Lincoln-Zephyr prow was sketched by Bob Gregorie in just 30 minutes

5. The streamline motif extends even to the boot handle

pioneering work on the self-starter before taking charge of R&D at GM's Dayton research laboratories. Kettering rejected Tjaarda's design for a small capacity, high speed engine, but, as with Earl, their paths were to cross later on. 'Boss' Ket's famous dictum – 'My interest is in the future because I'm going to spend the rest of my life there' – would have struck a chord with the ambitious young Hollander.

It was during a brief subsequent stint at Duesenberg, where he is said to have worked on bodies for the Model X in the winter of 1925, that Tjaarda was recruited as a design consultant by Locke & Co, coachbuilders of Rochester, in upstate New York. It was a large concern; in addition to supplying mainstream manufacturers and custom coachwork for Lincoln, the company made bodies for LeBaron and the Minerva agent in New York, Paul Ostruck, as well as for Frank deCausse, who had been poached by Locomobile from Kellner et Fils in Paris. Tjaarda's contract with Locke allowed him to continue developing the Sterkenburg prototypes, and in 1926 the independent syndicate he formed sought backing for his third car, the C3.

This small four-seater was powered by a mid-mounted sleeve valve engine of Tjaarda's own design, with a three-speed gearbox aft of the rear axle. It was built around a Lambda-like hull fabricated from steel pressings, with a structural dashboard. All-round independent suspension was by parallel arms, with rubber in torsion as the medium. The brake shoes operated directly on the inside of the wheel rims, and the engine, transmission, radiator, battery and rear suspension were combined in a single self-contained unit attached at only three points for rapid removal. He chose to clad these avant-garde underpinnings in a deliberately conventional-looking, snub-nosed, square-rigged fabric body as reassuringly pedestrian in appearance as Hans Nibel's W17 Mercedes-Benz of five years later.

Despite this, there were no takers, so Tjaarda abandoned this uncharacteristic reserve with the subsequent Sterkenburg C4s of 1931, comprising a stretched version of the C3 and a radically streamlined sedan that bore a striking resemblance to Edmond Moglia's mid-engined V-8 aerodynamic saloon for Prince Djelaleddin of 1930. With a modest dorsal fin presaging the Type 77 Tatra, the C4 streamliner echoed progressive European design thinking: a low, fully integrated pressed steel body and frame, bifurcated like a wishbone at the rear to accommodate the power module, independently suspended by rubber all round, with Tjaarda's own ohv twin-carb aluminium V-8 mounted amidships, teardrop wings (fully enclosed at the back) and a short, rounded grille-less nose.

According to some accounts, Tjaarda was recruited by Harley Earl's influential new Art and Color section at GM in Detroit 1930, but this can only have been as a consultant. What is certain is that the nationwide publicity generated by the dramatic C4 streamliner drew the attention of Walter Briggs, of the Briggs Manufacturing Co. A major supplier to Ford since the days of the Model T and Detroit's largest independent body producer, Briggs had pioneered low cost four-door saloon bodies in the US with its revolutionary coach body for Essex in 1922. It had then acquired LeBaron in 1926. By 1931, however, the economic malaise was biting hard, even for Ford. What's more, Dearborn's portfolio was conspicuously unbalanced – GM had no fewer than four marques attracting paying customers between the top-of-the-range Ford at $660 and the cheapest Lincoln, the KA, which cost five times as much. Henry Ford had bought Lincoln (for $8 million) in 1922 'to give "the boy" [Edsel] something to do'. However, he had rejected as superfluous all the refinements that Edsel wanted to build into the product. "Old Henry Ford had absolutely no interest in the design and appearance of the car," conceded Bob Gregorie. "He had a mania for simplicity, plain simplicity." At a time when styling was coming to the fore with Harley Earl's butch streamlining and heavy chromium appliqué, Walter Briggs recognised the imperative to help modernise his main client's output, if only for reasons of self-interest: there was talk of Ford closing down Lincoln altogether.

He instructed Ralph Roberts, principal of LeBaron, to entice Tjaarda away from his GM and Locke commitments, and he was duly hired to run Briggs's new body design and engineering department in 1932. In strict secrecy (Briggs was working on Carl Breer's Chrysler Airflow at the time), Tjaarda soon drew up an evolution of the C4 streamliner, a full-scale wooden maquette of which so impressed Edsel Ford that he asked his dictatorial father to commission a prototype for inclusion in the company's Exhibition of Progess roadshow that toured the US in 1933 as 'A designer's concept for a car of the future'. It also starred in Chicago's Century of Progress exhibition the following year, with the caveat 'This car is our design contribution and conception of a streamlined automobile and has no reference to current or future Ford production.' Nevertheless – and surprisingly for such a daringly unorthodox form – Ford's market researchers established that four out of five visitors liked the shape. It was enough for Briggs to give the green light to build two more: a front-engined variant to test the strength and durability of the body engineering, and a rear-engined car with a one-off alloy V-8 and automatic transmission, in line with Tjaarda's original intention.

But Ford was no Preston Tucker. An engine in the back was a step too far – in terms of design philosophy, as much as production implications. But Edsel was convinced by the rest of the concept, most notably its truss-frame unit construction, whose side members extended up the A pillars, over the doors and down the C pillars, allowing an exceptionally shallow floor pan and chairhigh seating. 'Beneath this outward beauty is a framework of steel trusses – the famous "arch bridge" construction,' cooed the advertising copy for the new model's launch. 'You have stood on such a bridge and marvelled that so light a structure could support great weight. Here is the same principle – body and frame a rigid unit combining lightness and strength.'

Edsel asked Briggs to have Tjaarda design a suitable new front end, but he was loth to comply. After some weeks, Edsel himself visualised a solution with sharp, vertical lines and instructed Gregorie to "help Tjaarda out". Trained as a yacht designer, Gregorie was familiar with the European design idiom from a family stable that included Delages, Amilcars and Citroëns; three years previously, he had created Dagenham's pretty Model Y. An awkward encounter followed, in which Tjaarda eventually acquiesced. Gregorie sketched his inverted boat hull of a bonnet on the back of a blueprint, extending the line forward horizontally, curving down to a near-vertical leading edge. It took him 30 minutes. The job cemented Gregorie's reputation with Edsel and paved the way for his later becoming Ford's first design chief.

Gregorie also substituted front-hinged doors for the Sterkenberg's four rear-hinged ones. In deference to his client's predilections, Tjaarda had already replaced his rubber independent suspension with beam axles suspended by transverse semi-elliptics. Antediluvian as this arrangement may have been, the unsprung weight was less than with GM's Dubonnet system.

Despite his father famously having "no use for an engine that has more spark plugs than a cow has teats," Edsel sanctioned a new 75-degree V-12 developing some 110hp – 25 more than the Ford flathead V-8 on which it was loosely based, but without the exhaustive development a Ford unit would have had. As a result, its thermal shortcomings were even worse. In retrospect, Gregorie was dismissive: "That engine was rough. It had a whip in it at certain speeds – vibration periods. Beautifully built, but not a long-life engine. Not a silky smooth engine as a 12 should be." Compared to Tjaarda's innovative original, the production car was undoubtedly a compromise. GM designer David Holls summed it up thus: "The Airflow was an engineering masterpiece, without the help of a designer; the Lincoln-Zephyr was a masterpiece of styling and design with an absolute lack of engineering below the surface.

"I think the car probably would have been more appropriate with the rear engine and sheep's-nose front end, as Tjaarda had originally conceived it," Gregorie himself later admitted. "The Zephyr had poor weight distribution. It was too nose heavy, and it was too light in the rear end. The Zephyr was beautiful to look at and was beautifully finished, but it wasn't the best piece of engineering in the world."

None of this, however, prevented the car from being a commercial success at a critical time for the company. At around $1300, it was keenly priced. It weighed 1520kg and, without the benefit of any wind tunnel work, boasted a creditable Cd of 0.45. And it was undeniably sexy, a car to be seen in. The name itself was of course a nod to Budd's massively influential Burlington Zephyr diesel electric streamlined train, a gleaming stainless steel symbol of a clean, bright technological future for everyman. The Lincoln-Zephyr went into production at Briggs in June, 1935 (with engine and drive train installed at the Lincoln plant), and was launched that November against Chrysler's second generation Airflow Eight. It was the most radical product Ford had ever attempted; only the Cord 810 threatened to steal its limelight. By the time the last Lincoln-Zephyr rolled off the line in February, 1942, some 133,000 had been sold.

H36109: *The Automobile*'s Lincoln-Zephyr

Regular readers will know of *The Automobile*'s Oily Rag Collection, a growing assortment of cars in various states of decrepitude but all highly original. One of the best preserved, surviving in extraordinarily untouched condition, is a 1937 Lincoln-Zephyr in right-hand drive configuration.

Lincolns in the UK were normally imported by Lincoln Cars Ltd, a wholly owned Ford Motor Co subsidiary with premises on the Great West Road at Brentford. Only the larger Ford distributors would have kept a V-12 Lincoln-Zephyr in stock; the smaller dealerships would order direct from Lincoln Cars when required.

Our car, chassis/engine number H36109 and registration DVU 170, was delivered new on 1st October, 1937, by Gordon's of Heaton Chapel, Stockport, Manchester, to one Ernest Berry of Wellington Road, Heaton Chapel. It cost him £495 and was ordered without any accessories – not even a heater. The original colour was Mercury, a light green semi-metallic shade that still remains on the dashboard. The rest of the car was repainted, possibly before the war, no doubt because the original cellulose, like all early metallic paints, had begun to flake.

It would be interesting to know more about Mr Berry, who appears to have been a solicitor. Whoever he was, he appears to have kept his Lincoln-Zephyr until at least 1962 – possibly longer, but that is when the last tax disc expired before the car was laid up. He used it sparingly, and never in winter. According to the log book, it was taxed for summer motoring only, which is presumably why it has survived in such extraordinary condition. Even known rust hot-spots for the model are still as they left the factory.

The car was sold in 1974 to Francesco Mirabile, an Italian living in Brussels. He drove it onto the Harwich-Hook of Holland ferry on 14th June that year, paying import tax of 891 Belgian francs and VAT (BTW) of 2310 francs on a declared value of 8100 francs. He did some conservation work before retiring to Italy, whereupon he sold it to his friend Jacques Deneef, another Brussels-based collector. Much later, Signor Mirabile sent Colin Spong, known as England's Mr Lincoln-Zephyr and the man we bought the car from, the original rear number plate which had been hung on his garage wall.

Since it joined our collection in the spring of 2011, we have used the Lincoln extensively. Within a week of its arrival at our Cranleigh offices, we had driven it to Wales and back to take part in the second Oily Rag Run; during the 400-mile round trip it didn't miss a beat. Later that summer, it was an invited entry in the prestigious Cartier Style et Luxe concours at the Goodwood Festival of Speed, sharing a streamlining class with a Tatra T87 and Norman Foster's Dymaxion recreation, among others.

This year, we have upgraded the car at the suggestion of Colin Spong and his brother, Adrian, both of whom are gurus of all things Lincoln-Zephyr. They sourced and fitted a correct Columbia two-speed axle, a period accessory which improves immensely the usability of this otherwise low-geared car. Giving, in effect, six forward speeds, it most importantly offers much more relaxed cruising in top – as well as improving fuel consumption on long journeys.

We have been using the car more this past autumn, including an expedition to north-west Essex for our third Oily Rag Run, where we chauffeured around Nick Clements of *Men's File* magazine and his friend Annika Caswell, who appears in the opening picture, taken by Nick, of this article.

The Lincoln-Zephyr's gorgeous looks and smooth V-12 engine attract admirers wherever we take it, and driving it is a pleasure. Although the 4.4-litre engine is willing to pull from low speeds in high gear, we try to give the V-12 the revs it needs to avoid the problems experienced by prewar owners, who failed to give their engine the exercise it required. The brakes, gearchange and especially the handling are all commensurate with their Ford V8 ancestry – that is to say, adequate. But for long-distance, unhurried travel, the Lincoln provides unbeatable comfort and style for four, five or even six passengers.

Many cars are legendary – almost anything made by Bugatti, Hispano-Suiza and Duesenberg, for example, and perhaps even the Austin Seven – but few are so rare and so rarely seen that they have become beasts as mythical as the Chimera and the Unicorn. Among only a handful of British rarities, what has long been known as the Embiricos Bentley is now all but unknown in the metal and we speak of it with awe, supposing it to have done great things, but knowing only that it was 'streamlined' – in the Thirties the common word for aerodynamics – at a time when most Bentleys of its day, all bodied by coachbuilders, were exquisite examples of traditions that reached back to the carriages of the 18th century. Having neither seen nor heard of it since its last appearances at Le Mans in 1949, 1950 and 1951, and now into my 80s, I have just seen it again, in the spotless Bentley factory in Crewe, exhibited with an R-Type Continental, the son it is supposed to have spawned in 1952.

As the Chimera was a monster with a lion's head, a goat's body and a serpent's tail, it is a not unfair description of this serious attempt to produce a full four-seater aerodynamic sports saloon in 1938 – to our eyes too tall and narrow, and from some angles gauche. It is, nevertheless, exquisite in detail, every junction considered so as to smooth the line from nose to bonnet, bonnet to split screen and roof, and the roof tapering through the boot lid to a teardrop point to fit the form as much as the line; most important, the proud Bentley radiator is replaced by a curved and sloping cowl, the headlamps faired between it and the wings. Even so, there is, overall, a sense of disjuncture in the whole, the flat glass of the cabin jarring against the curvaceous metal, the spatted rear wheels a heavy and unrelated afterthought, the lion's head and the serpent's tail far the best parts of it.

This Bentley was the consequence of a conspiracy. Early in 1936, when all over Europe and America manufacturers had already attempted clumsy, largely infelicitous and almost wholly unscientific streamlining by endowing their cars with rearward-leaning radiators and humped backs to give the impression of improved performance, two directors of Rolls-Royce opined that 'The technical advantages of streamlining are so great that the adoption of the principle is inevitable…It is, of course, accepted that our clientele is probably the most conservative in the world…' Was this so? Had they forgotten Gurney Nutting's rakish coupé of 1930 for Woolf Barnato on a 6½ Litre chassis (that is before the Rolls-Royce takeover)? Or the fastback saloon body concocted for a 4½ Litre from one designed by Walter Belgrove for the Triumph Gloria Vitesse of 1934 (the post-war Roadster and TR2 were also his designs)? Its tapered tail was a remarkable anticipation of the Embiricos car.

Walter Sleator, the Paris agent for Rolls-Royce, then observed that their cars were always being overtaken by the Bugattis, Delahayes and Talbots of the time. W A Robotham at Derby opined that they were 'about as bad

In the 1930s, this was every schoolboy's idea of a 'streamlined' rear end

The Embiricos Bentley

1. Unlike so many other so-called aerodynes of its era, the Embiricos Bentley was actually tested in a wind tunnel in this wooden mock-up form

2. Brian Sewell, the London *Evening Standard*'s scourge of the contemporary art 'mafia', focusses his steady gaze on the Bentley's frontal contours

as any we know for lack of streamlining'; he then performed a wind tunnel test indicating that the performance of most coachbuilt Bentleys would improve by 15mph if they were driven backwards. For this he blamed the lofty radiator, but for Bentley themselves to dispense with it was quite unthinkable. There had been earlier experiences in streamlining by British manufacturers and coachbuilders – Rover, retaining its original vertical radiator, had produced three humpbacked saloons, and Thrupp and Maberley had, in 1935, built a four-door saloon with a particularly elegant sloping tail, improving the kindred line of a Park Ward car of the year before, but both were fronted by the classic radiator and suffered the British disease of too much weight and too little power to make any difference to the performance. For the Bentley nose to go, a Frenchman must take the blame.

For my own part I am quite sure that, given a clear brief to abandon the classic Bentley radiator, Gurney Nutting, Corsica, or the back street Corinthian, could have made an efficient aerodynamic body, but perhaps the directors of Bentley did not want even a whisper of such a thing to be heard by their supposedly conservative clientele. The task went, indirectly, to the French aerodynamic engineer Georges Paulin, who was in a partnership of sorts with one of the smaller Parisian

1. Even if certain details do offend our critic, most enthusiasts would rank the Embiricos coupé among the prettiest cars of the late 1930s

2. The finished car cost André Embiricos almost £6000, more than twice the price of the most expensive Rolls-Royce Phantom III

The Embiricos Bentley

coachbuilders, Marcel Pourtout. He was already at work on useful designs on Peugeot chassis, and had produced, in 1937, an aerodynamic coupé on a Delage chassis from which the general form and some details of the new Bentley were derived. The commission was indirect because neither it nor payment came from Derby; instead, Walter Sleator was to persuade a French client to buy a Bentley 4¼ Litre chassis and to commission and pay for an aerodynamic body. Thus, if the car proved to be a disaster, Bentley themselves could claim that they had no responsibility for it.

Sleator's chosen client was André Embiricos, a wealthy Greek banker resident in Paris. In December, 1937, Rolls-Royce charged him the full £1150 for the chassis. By the end of January, 1938, a scale wooden model had been tested in a wind tunnel. In March the chassis arrived from Derby, and by July the body, in lightweight Duralumin, weighing some 150kg less than a conventional Bentley, had been mated to it and the interior dressed, the complete car ready for road testing; on the 18th of that month, Bentley guaranteed the car. Finished, the car cost Embiricos almost £6000, more than twice the price of the most expensive Rolls-Royce Phantom III; he took delivery on 8th September. It then circled Montlhéry for an hour at an average speed of 110mph. In the early months of 1939 it made a press trip into Germany to see what it could do on the new Autobahns and, shipped to England, George Eyston drove it round Brooklands for an hour at an average speed of 114.63mph. In essence the chassis and engine were those of the 4¼ Litre Bentley, the marque's only model since 1935, but the engine was given larger carburetters and modified to give 142bhp instead of 125, and a higher axle ratio enabled the car to cruise at 110mph at 3500rpm, with a theoretical maximum of 120mph that it never quite reached; the suspension too was tweaked to improve roadholding, but without the authority of Rolls-Royce – for which impertinence Walter Sleator almost lost his job.

Poor Embiricos – he seems hardly to have taken the wheel of this extraordinary machine and in July, 1939, some five weeks before the outbreak of World War Two, sold it to Soltan Hay, the owner-driver who drove it on family holidays and put it through its paces in its post-war appearances at Le Mans. In 1949 it was sixth overall – a wonderful achievement, it was said, for an 11 year old car, but in six of those years the European industry had been dormant, if not destroyed, and in the last four Bentley had relied on the revival of a prewar engine and chassis little changed, the German industry had

1. Stripped of its heavy rear wheel spats and bumpers, the decade-old Embiricos Bentley looks every bit the racer at Le Mans post-war. The vents in the rear wings, for brake cooling, are still there today, as are corresponding frontal inlets

2. Whilst the Bentley works affected to take no interest in the car, it was tested extensively throughout Europe

scarcely recovered, and the French, with the beautiful *Grandes Routières* that had been the Bentley's inspiration, had been all but taxed out of existence. The Embiricos car, unused for the duration of the war, was in essence only five years old and should have done much better; was it really such a great design?

Many of the *Grandes Routières* were so perfectly conceived and executed that they make the Embiricos Bentley look like a one-man garage job; the formula is constant – the nose still high and assertive even though it is not the classic vertical that gives a car identity, the headlamps faired into front wings that may bulge to enclose the wheels, the rear wings spatted, boots tapering to nothing and crammed with spare wheel and petrol tank that deny every cubic inch to luggage, cabins comfortable for two but unbearably cramped for four, the sloping roof too low. It must be admitted that Letourneur et Marchand produced prettier cars for Delage, that Figoni et Falaschi again and again produced perfection for Delahaye and, particularly, for Talbot-Lago (a marque for which Pourtout and Paulin worked with a very heavy hand the following year), and that the Bugatti 57SC Atlantic, designed in house, is the most extreme and extravagant (most beautiful or hideous?) of all these teardrop coupés. What, I wonder, would the flamboyant Iakov Saoutchik have conceived for the Embiricos Bentley?

The teardrop was the genre of the day in Parisian aerodynamics, the rounded nose bludgeoning the air with brute force, the tapering body slipping easily into the space made for it – this was the intuitive thinking of the Art Deco Thirties and no one in France paid much attention to the middle Europeans Rumpler, Ledwinka, Jaray and Kamm, whose theories were much nearer the truth. In 1921 Paul Jaray was first to begin wind tunnel tests on model car bodies, proving that the true teardrop tail must be a teardrop in plan as well as elevation – that is, a teardrop in every aspect and not just in profile; to achieve this, the best place for the engine is in the rear, the cylinders in a V to reduce its length; reductions in drag could be achieved by concealing the springs and other excrecances of the chassis, by sinking door handles into the door (as in the Embiricos Bentley), and by shaping sidelights and other elements as teardrops too. At the Paris Motor Show of October, 1933, '*Carrosserie Aérodynamique*' was everywhere, but only as variations on the teardrop seen in profile; that same month, at the London Show, a

The Embiricos Bentley

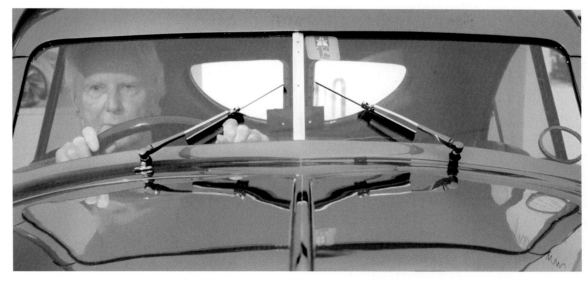

As Brian Sewell suggests, these thick windscreen pillars contrast with the ultra-thin equivalents on Paulin's Delage D8-120S of a year earlier

number of close-coupled coupés, including a Park Ward Bentley, toyed with the idea of the sloping roof and tail, but not with the sloping radiator. In the Berlin Show of March, 1934, Maybach showed its huge eight-litre Zeppelin chassis with a limousine body influenced by Jaray – full-width (no running-boards), an uninterrupted wing-line rising at the front and falling to the rear in almost equal measure, lamps faired into the wings, and, most revolutionary, a sloping split windscreen and a radiator not only sloping but describing an inward rather than an outward curve. Tatra was even more daring with the world's first serially manufactured and genuinely aerodynamic rear-engined car, the T77 – a six-seater luxury saloon with exceptionally low drag, powered by an engine of only 59bhp that gave a top speed just short of 90mph, all but matching that of any Bentley with a conventional coachbuilt saloon body but using only half the power. If any car of the day proved the virtue of aerodynamics, it was this Tatra. In 1935 another Jaray Maybach (on the SW35 chassis) with a deeply curved windscreen and headlamps in the wings, its streamlining exaggerated by two-tone paint, caused uproar at the Berlin Show; it anticipated not only the Embiricos Bentley but the Citroën DS.

By then almost every French marque had to some degree adopted streamlining for cars both large and small, cheap and expensive, but not so in England. It was 1939 before snub noses (but little else) had been adopted by mass manufacturers, but Bentley retained its classic radiator and its coachbuilders continued to refine their classic idioms, now so fastidious in line and detail, especially when razor-edged, that the mannerisms of aerodynamics seem to have sent them scurrying back to their design studios to polish and perfect their ancient traditions and eschew the future. They were, however, comfortable as well as elegant – and comfort was not, for me, a characteristic of the Embiricos Bentley. I imagine it was hell to drive; one sits low, peering through the narrow screen, the long high bonnet obscuring what is ahead, lateral visibility wretched, everything to the rear invisible, other than two small patches of sky through the paired windows, and, worst of all, the driver's seat not aligned with the steering wheel so one must sit askew. Early drivers complained that the cabin was too hot and access to the rear is adequate only for small dogs. Bentley were to discover that without the wind resistance of a traditional body, the brakes must be mightily improved.

The Embiricos Bentley attracted a handful of orders for replicas, but these were ignored – the panjandrums of Derby had in mind the better idea (in terms of sales) of a full four-door saloon 'with plenty of luggage space'. They turned direct to Paulin, suggesting in July, 1938 (a date proving how swiftly they responded favourably to the Embiricos car), that he might find inspiration for a new Bentley nose in that of the Lincoln-Zephyr – and he did. This was a prototype Corniche intended for production in 1940, the year in which a stray German bomb destroyed it while on the quay at Dieppe awaiting ship to England. By then Derby had misgivings that the body, constructed by Vanvooren to Paulin's drawings (Vanvooren was a rather larger Parisian coachbuilder with established connections with Rolls-Royce as well as Pourtout and Paulin), was more American than English in character. It was, perhaps inevitably, another Chimera, part Lincoln, part Embiricos, its extended tail later as conventionally booted as all the post-war coachbuilt bodies on the Mark VI chassis.

There was no further collaboration between Bentley and Paulin. Four incomplete Corniches lay, discarded, at the Rolls-Royce depot in Châteauroux; Vanvooren worked mainly for Delahaye after the war and went bankrupt in 1950. Pourtout too worked for Delahaye in 1946, but then turned to commercial vehicles. Georges Paulin, who had come to think of himself as 'a small part of the Rolls-Royce family', joined the French Resistance in September, 1940, was betrayed, arrested in November, 1941, tortured by the Nazis and executed on 21st March, 1942; he was 40. Bentley claim that the genes of the Embiricos car are evident in the R-Type Continental of 1953, on the fastback of which James Bond's conjugation with a maiden at such an angle is so intriguing (never translated into film, it remains a literary conceit). I am inclined to argue that this paternity is not so and that this much admired car is as American in origin as the aborted Corniche – it is the fastidious refinement of the five-passenger Club Coupé or Sedanette that was by 1946 common in General Motors from Chevrolet to Cadillac, but still fronted by the classic ancestral radiator of the Bentley.

Paulin deserves a monument in England. Three-quarters of a century on he is unlikely to get it. We could make some small amend by dubbing the Embiricos car 'The Paulin Bentley', as Rolls-Royce did themselves. Embiricos paid with money, Paulin with his life.

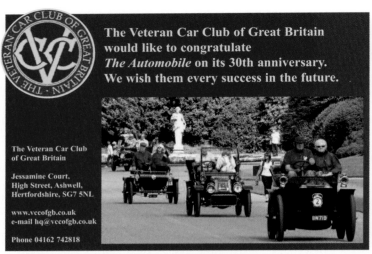

The Veteran Car Club of Great Britain would like to congratulate *The Automobile* on its 30th anniversary. We wish them every success in the future.

The Veteran Car Club of Great Britain
Jessamine Court,
High Street, Ashwell,
Hertfordshire, SG7 5NL

www.vccofgb.co.uk
e-mail hq@vccofgb.co.uk

Phone 04162 742818

FRICTION SHOCK ABSORBERS

COMPLETE RANGE OF NEW UNITS & SPARES

Andre Hartford Limited
Upper Battlefield, Shrewsbury SY4 3DB
Telephone Jamie Dowley on:
01939 – 210458   Fax: **01939 – 210644**

SERDI UK

- Specialist engineering services – machining; grinding; boring; balancing
- Supply & fit of key components – pistons, conrods, valve guides & seats
- Complete assembly of major components
- Extensive experience of vintage & historic engines.

SERDI UK Ltd
Tel: 01895 232215 Fax: 01895 253667 Email: sales@serdi.co.uk **www.serdi.co.uk**

DONNELLY UK LIMITED

ALLIED RUBBER PRODUCTS
P.J. DONNELLY RUBBER

Mouldings and Extrusions in natural and Synthetic Rubber

■ PRECISION MOULDINGS TO CLIENT SPECIFICATIONS
■ EXTRUSIONS IN THERMOSET AND THERMOPLASTICS

Tel: 0121 565 0988 • 0121 565 0961
Fax: 0121 565 0976

Unit 15, Cornwall Road industrial Park, Smethwick, Warley 66 2JT

Spares Manufactured and supplied for all
Overhead Camshaft models 1929-36
Comprehensive stocks

New ninth edition catalogue now available £10.
Upper Battlefield, Shrewsbury SY4 3DB
Tel: 01939 210458   Fax: 01939 210644

Veteran, Vintage and Classic Restorations and Repairs

Sunbeam, Talbot, Darracq
and many other Marques.
Full chassis up restoration
including Engine, Bodywork,
Electrics and Trimming
Pre MoT servicing. No job too small.
Easy access from M6 - M54

Contact: Stuart LLoyd
Tel/Fax Day - 01902 737008
Evening - 01902 734141

Congratulations to *The Automobile*
on their 30th Anniversary

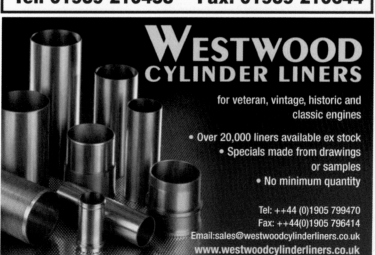

WESTWOOD CYLINDER LINERS

for veteran, vintage, historic and classic engines

- Over 20,000 liners available ex stock
- Specials made from drawings or samples
- No minimum quantity

Tel: ++44 (0)1905 799470
Fax: ++44(0)1905 796414
Email:sales@westwoodcylinderliners.co.uk
www.westwoodcylinderliners.co.uk

Country Lane Tours

'It is not the cheaper things that we want to possess, but expensive things that cost a lot less' – John Ruskin

2013 EVENTS

3rd CROATIAN CRAWL, May 19th – June 4th
(**FREE** DVD of 2008 tour available on request). Cars and crews are whisked across Europe by overnight Motorail sharing a route along the edge of the River Rhine used by the famous Orient Express.

Leaving the train in Villach, Austria, cars travel via Ljubljana in Slovenia to Opatija & Pula on the Istria Peninsula, before island hopping to Zadar on the Dalmation Coast and hugging the sea edge to Split and Dubrovnik.

The return run to the train follows the Croatia/Bosnia border through the mountains – brilliant driving roads and spectacular scenery!

WELSH WEEKEND, APRIL 26th – 29th (30th).
Ideal to brush the winter cobwebs of your car the event comprises of two days touring on the beautiful Isle of Anglesey, using quiet lanes and visiting numerous attractions. You'll be amazed what Wales has to offer!

2nd DRAGON TRAIL, August 16th – 23rd.
The first Dragon Trail was met with such acclaim by participants it's running again in exactly the same format. Starting with three days in South Wales, staying at a five-star hotel, the route winds north via Hay-on-Wye, the Victorian spa town of Llandrindod Wells and Llangollen in the Dee Valley to final overnight halt and gala dinner in the famous Italianate village of Portmeirion, nestling on the edge of the Mawddach Estuary.

SWISS SWIRL, September 18th – 29th.
We meander through rural France via Beaune to Interlaken where we take in all the surrounding mountain passes including the old St. Gothard (still cobbled!) and, of course, a trip on the lake in the paddle steamer.

Heading towards home we take in the Schlumpf Bugatti collection in Mulhouse, picturesque Epinal, the old Reims racing circuit and guided tour of Mumm's Champagne.

2014 events: Welsh Weekend, April - Viking Tour (Sweden & Norway), May - Dragon Trail, August - Unknown Italy, September.

All tours are designed to be a holiday with your vintage or classic car, not a mad dash with different hotels each evening.

Routes are arranged to allow plenty of time for sightseeing and with numerous two-night stops you are able to relax without the stress of having to pack bags and move on each morning.

For peace of mind all continental events are accompanied by a service vehicle crewed by qualified mechanics and an experienced tour director is always available to deal with any queries. In addition, a detailed road book and marked map make certain you don't get lost.

For comprehensive details of all events email your name and postal address to countrylanetours@aol.com or call Ian on 01824 790280.

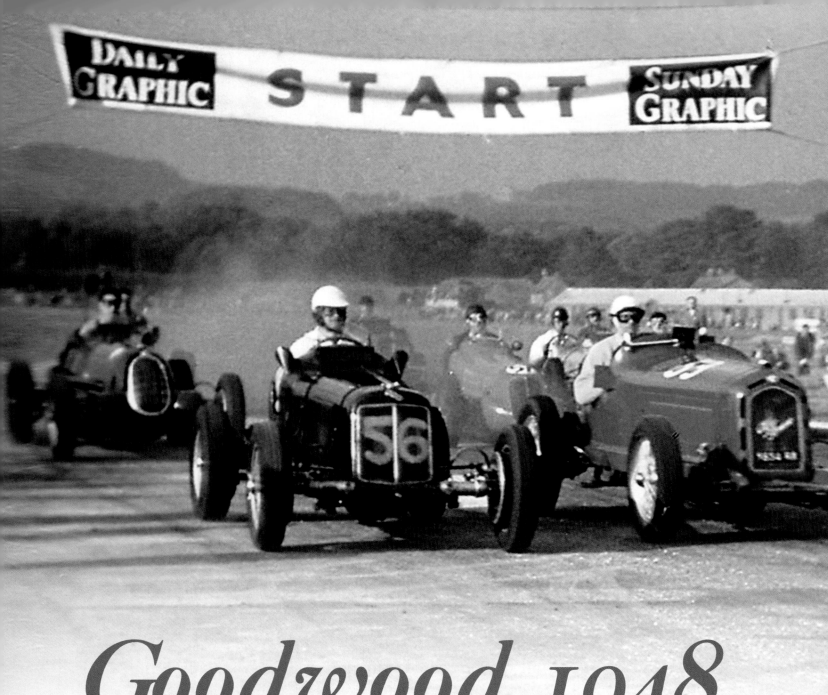

Goodwood 1948
The first meeting

As a schoolboy **David Venables** attended the first meeting at the new Goodwood circuit. Here he recalls that September day of 64 years ago

The start of the over 1450cc racing car race. George Nixon in ERA R2A (No 56) and David Lewis's Monza Alfa (No 53) lead, with Geoffrey Ansell (ERA R9B) coming up on the right. Dennis Poore (8C-35 Alfa), who won the race, is behind Nixon (*The Grand Prix Library*)

When Goodwood opened for the first meeting in September, 1948, I was a teenager, living with my mother in a flat on Kingston Hill. Not far away in the same block was Edgar Kehoe. Edgar had raced at Brooklands in the '20s and '30s and had driven a Riley in the 1930 Double-12 and the 1932 1000 Mile race. He had a Brooklands Riley which he was racing in post-war events. It was kept in a lock-up garage behind the flats and he let me help him work on the car. Edgar's wife Mimi was the secretary to Archie Frazer-Nash, who lived about 100 yards further up Kingston Hill, while Archie's mother Dr Lilias Frazer-Nash lived in the flat above us. My mother was a friend of Dr Frazer-Nash and of Archie and his wife.

Edgar had entered the Riley for the opening Goodwood meeting. There was great excitement that a permanent racing circuit was coming at last, after British racing had been starved of a circuit for the first three post-war seasons. I had to be there as I knew motor racing history would be made. Edgar drove down to Goodwood on the day before the meeting for practice, accompanied by Mimi. Archie Frazer-Nash was a steward at the meeting and, to my great joy, agreed to take me to Goodwood on race day. We travelled down in his 1938 4¼ Litre Park Ward Bentley saloon, accompanied by Elisabeth Quirke, wife of Paddy Quirke, a close friend of Edgar. Paddy was a keen 'Nash owner who had been in the RAF during WW2 and had been a prisoner of war in Colditz with Tony Rolt. He was to come to Goodwood later in his 'Nash

1. Stirling Moss, followed by Eric Brandon, line up to return to the paddock in their Coopers after the 500cc race. Curly Dryden in Cooper No 10 seems to be starting another lap (*The Grand Prix Library*)

2. The saloons at the start of the first race. Ray Brock's HRG No 7 leads from Patrick Hall's Healey Elliot with Pycroft's winning Jaguar on the outside following Hall (*The Grand Prix Library*)

accompanied by George Booth, a local solicitor who also helped in preparing the Riley. The sister of the very glamorous Elisabeth, Paddy Naismith, had raced at Brooklands in the early 1930s.

It was a rapid run from Kingston to Goodwood, as Archie's racing experience showed in his road driving. During the journey, he asked: "Are we in a 30mph limit?" I said I believed we were and he commented, "We must get out of it as quickly as possible." When we arrived at Goodwood we drove into the circuit by the main gate, now the principal entrance for spectators.

Archie parked the Bentley beside some ex-RAF huts which are still there and, as he climbed out, said "There's George Eyston" and marched off to have a conversation, taking Elisabeth with him. I had to find a way across to the paddock, but I paused for a moment to look at the D6-70 Delage which had won the 1938 TT and which was parked nearby.

I went to the paddock crossing, which is in the same place now. It was being guarded by an elderly man. I lacked a paddock pass, but I explained I had an urgent message for a driver and needed to speak to him at once.

Goodwood 1948—The first meeting

3. Prince Bira talks to Peter Walker, who leans on ERA R7B (*The Grand Prix Library*)

The man smiled and waved me across the track. It doesn't work like that at Goodwood these days. The paddock adjoined the circuit, on its present site; it was surrounded by a chestnut paling fence and separated from the track by about 10 feet of grass verge. The cars were lined up in a row down each side with wire mesh laid down the centre, of the type used to make temporary wartime air strips. The cars were grouped in classes, with the sports cars on the track side, facing the racing cars.

I walked down the line of sports cars to the 1100cc class. Edgar's Riley was the last car in the class, parked between the HRG of Joe Lowrey, *The Motor*'s Technical Editor, and the SS100 coupé of Paul Pycroft, which was running in the saloon car race. The Riley appeared to need no work, so I accompanied Edgar on a walk round the paddock. The star turn was the shining new 4CLT/48 Maserati of Reg Parnell. At that time it was the very latest GP car, and I was especially taken by the Borrani wheels.

There was much excitement about a new driver, Stirling Moss, who had been breaking records in hill climbs and sprints in a 500cc Cooper; this was to be his second circuit race. Moss's cream-painted Cooper was lined up with the 500s and it was noticeable that the Coopers in this class made the other runners look rather amateurish. The paddock was primitive, and the gents was a hole dug in the ground surrounded by sacking screens.

I had been an avid reader of Prince Chula's book about Bira's racing. At that time Bira was a rather exotic and legendary figure to me, so I was fascinated to find him wandering round the paddock talking to the drivers. The opening meeting was run with a National licence, so Bira was unable to compete, but he had flown in with his twin-engined Miles Gemini finished in Siamese blue and yellow and this was parked outside the paddock fence. The Duke of Richmond was also walking round the paddock. Edgar had been at school with him and raced against him at Brooklands so, with much laughter, made some teasing comments to the effect that the Duke should be racing, not presiding over the meeting. Soon after this His Grace drove round the circuit in his Bristol 400 for its formal opening.

The saloons came out for the inaugural race. A week earlier, *The Motor* had said the saloon field would comprise six Healeys, a Jaguar and an HRG. As the cars came out I commented to George Booth it was probably an accurate prophesy of the finishing order. At that time the Elliott and Duncan-bodied Healeys were the must-have cars, claimed to be 'the world's fastest production saloons'. It all changed a month later when the first XK120 appeared at the Motor Show. The race didn't go according to form. Paul Pycroft's aerodynamic SS100, with a hard top, took off at the start of the three laps and ran away, leaving the Healeys holding off Brock's hardtop aerodynamic HRG.

Next were the 1100 sports cars, and I had quite high hopes of Edgar getting a place. Harry Lester in his L/N-type MG was likely to be unbeatable on form, but the rest were of comparable performance. The flag fell and Lester led away, going on to win easily, but by the time the field reached Madgwick from the start Edgar was already well behind and the Riley was misfiring. He staggered round at the back of the field, while at the front Len Gibbs (Riley) and Peter Morgan (Morgan 4/4) fought for second place. Until 1952, when the chicane appeared, it was a straight run to the finish from Woodcote, with just a mild bend before the line. At the finish Gibbs was leading but seemed to lift off as he saw the flag and Morgan nipped past. Edgar came in and said Lowrey had overturned in front of him at Woodcote. The Riley's problem was evident: one of the SUs was flooding and had wetted the plugs.

Before we could investigate the problem, Lowrey appeared with a bandaged hand but otherwise unhurt, and the battered HRG was towed in. Lowrey said he wanted to drive it home, but the body had collapsed on the passenger's side so emergency repairs were needed. A hydraulic bottle jack came from the Riley's tool kit and was placed under the edge of the HRG's body. I was

Automobile 175

1. Reg Parnell in his new 4CLT/48 Maserati holds off Bob Gerard in ERA R14B at Woodcote in the Goodwood Trophy

2. The paddock scene at the opening meeting

3. During practice, Edgar Kehoe snoozes in the sun, lying between his Riley and Pycroft's Jaguar

4. Archie Frazer-Nash, on the left, talks to the Duke of Richmond (right)

5. Dickie Le Strange Metcalfe in his Fiat Balilla leads Nigel Orlebar's Ford-based special out of Madgwick in the 1100cc sports car race

Goodwood 1948–The first meeting

told to pump it up until the normal position was reached, then Lowrey and Edgar wired a very large ring spanner to the chassis frame, which replaced the broken body support bracket. When the jack was let down the car seemed almost normal, apart from the dents and scratches.

While this was going on, the 1500cc sports cars had come out and raced. I gave the race occasional glances while the repairs continued and saw George Phillips win in his special MG TC from the HRGs of Ruddock and Meisl. The two-litre sports car race was expected to be a 328 BMW benefit. It was, but Ken Watkins beat the tipped Tony Crook after they had seen off the blown MGs of Jacobs and Leonard. The last car to finish was the Peugeot 402 of Dorothy Patten. Edgar had competed against her in prewar rallies and we had talked to her before the racing began. She was supposed to be rather glamorous, but I was already taking an active interest in such things and was disappointed.

There was keen anticipation for the 500s. It was the first time these had appeared on the big stage. A race at Gransden a year before had seen one finisher and there had been some short races at Brough, but there were still doubts if it would be a serious racing class. Eric Brandon in the factory Cooper had been the man to beat in sprints and hill climbs until the arrival of the young Moss, so a battle between them was expected.
It didn't work out. As the flag fell, Brandon stalled and was push-started way behind the field, which was led all the way by Moss. Brandon pulled up to second place by the end of the third and final lap. The race showed the 500 class had arrived.

In the immediate post-war seasons, Monaco Engineering at Watford was probably the leading British tuning shop. Bob Drew, a mechanic at Monaco, did some of the more exacting work on Edgar's Riley. While doing this he regaled me with stories of continental forays with Peter Monkhouse, who raced Dudley Folland's monoposto K3 MG. Folland was driving the K3 in the three-lap race for blown 1100s and unblown two-litre racers, so I was keen to see it in action. The grids for the opening meeting were balloted and Folland was on pole. He shot away at the start and was never headed. John Cooper in a 996cc Cooper-JAP held on to him on the opening lap, then fell away, and Frank Kennington moved up into second place in another K3, a long-tailed two-seater. There were four K3s in the race and I was entranced by their exhaust notes. On the last lap the Spikins Special went up into third place. This was an Amilcar C6 with a dirt-track Lea-Francis engine painted in a bright yellow and driven by Basil de Mattos. I spent much time during the next term at school making sketches of the Folland K3.

I had seen Dennis Poore win at Gransden the previous year in his 8C/35 Alfa Romeo; he looked a likely winner of the unlimited race which came next. He was near the back of the grid but came through at the start to take the lead, though harried by Peter Walker (ERA R7B) and John Bolster (ERA R11B). At the end of the second lap Poore had a big slide coming round the bend after Woodcote and ran onto the grass on the other side of the track from me. He can only have been a few feet from the unprotected spectators, but held the slide and pressed on. Walker was closing right up but lacked the power to get by, so Poore won and Bolster was third. This was the only time I saw Bolster in action. He gave up racing following his bad accident in the British Grand Prix three weeks later.

The final race of the day was the five-lap Goodwood Trophy. Parnell won but was chased hard by Bob Gerard (ERA R14B). Gerard was on Parnell's tail all the way but couldn't get by. At the end the difference was about a length. It was a splendid race. The crowd flocked across the track and thronged the paddock fence to look at the cars, while Bira took off in the Miles Gemini and made a low pass over the paddock before departing. Bob Drew appeared and asked about the Riley's problem. Edgar had concluded that the copper finger which shuts the needle valve in the float chamber had jammed against the top of the float, causing the flooding, but Bob took a quick look and said the float was punctured.

It was decided the offending float should be taken to the hut across the track near the control tower, which housed the bar and refreshments. It was hoped it could be boiled in water to release the trapped fuel. The party followed Bob and Edgar across to the bar. I was a tall lad, so no one said anything when a half pint of beer was pushed into my hand. There was a big group of drivers and helpers discussing the day. I was talking to Nigel Orlebar, who had raced his Ford-powered Orlebar in Edgar's race. I asked him about his future plans and he said he hoped to become the Allard of the Ford 10. It was the start of a friendship which lasted until Nigel died in 2004.

After a while I left the party and walked back across the track to the paddock. The spectators had gone and most of the competing cars had gone, too. It was about 6:30 and the evening sun shone across the airfield. Standing rather forlornly in the middle of the paddock was Basil de Lissa's MG. This was an offset single-seat N-type built by Bellevue Garage before WW2, and I studied it closely. It was an unblown 1100 and I felt was a very desirable car to race with few bothers, though it had been crudely hand painted in semi-matt black. It is still raced now but is very shiny, blown and of 1500cc.

Edgar and Bob returned with the boiled float; this was fitted to the Riley, which ran on four cylinders again. Mimi, Edgar's wife, said she and Elisabeth Quirke wanted to go back to Kingston with Archie Frazer-Nash, probably with dinner en route in mind: would I mind accompanying Edgar in the Riley? For me this was the end of a perfect day, returning home from the first meeting in a competing car. It had been truly memorable, though I look back slightly wistfully as everyone in our little party has gone long ago. Of the drivers, there is only a handful left...

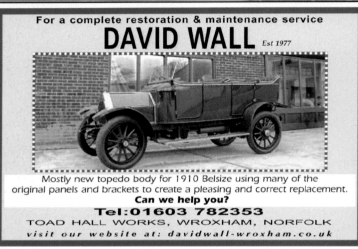

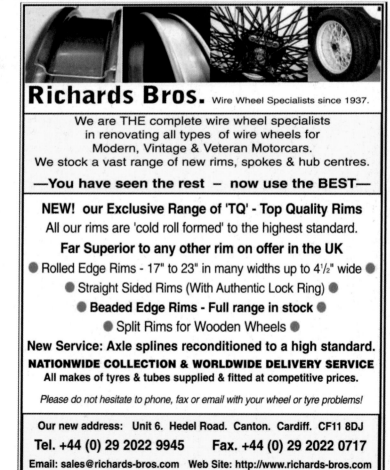

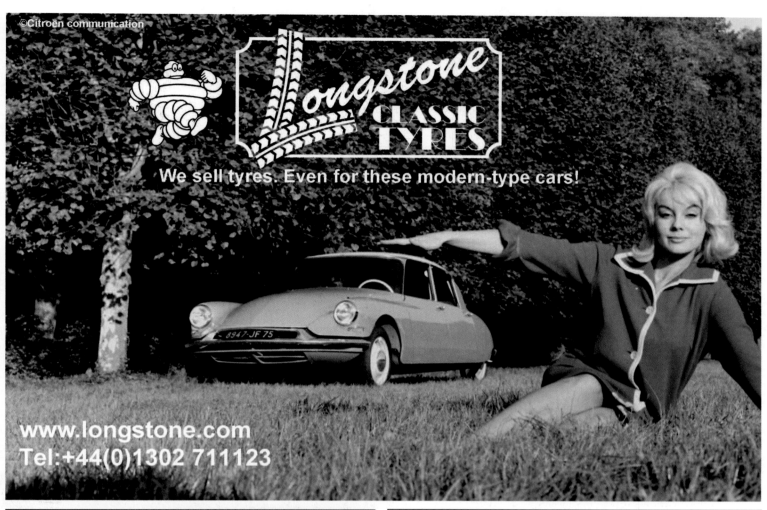

Elliptical Excellence

Saab's 92 prototype, the Ursaab, framed an often forgotten yet vital moment in automotive design. **Lance Cole** charts the details of the car and the man who shaped it

When the Saab 92 was launched in 1949 the motoring press freely admitted that, so unusual was its appearance, it was difficult to analyse its shape or pigeon-hole the design lineage. Even *The Autocar* was mystified by the new design themes that the little Saab saloon pioneered, suggesting that the shape resembled that of a small yet exotic sports coupé. Here was something new, a new design language that left the re-warmed pre-WW2 school of the late 1940s in limbo.

While other car manufacturers churned out tall, narrow cars with separate wing lines, multi-surfaced panels, running boards and upright windscreens, the strange Swedish car boasted a rakishly inclined windscreen angle, a domed roof with exquisite tumblehome and an aerodynamic shape that eased smoothly back into a tail section resembling the aftertip of an aircraft fuselage.

The production 92, Saab's first car, stemmed directly from the so-called Ursaab (original Saab, in Swedish) as its prototype. Seen from the present day, all Ursaab needs is the addition of an empennage and some chrome to look like a 1960s American 'space-age' design study from the studio of GM's Harley Earl, yet it was conceived in the early 1940s.

Ursaab was an integrated sculpture of utterly new form and one with a unique visage. In designer-speak, it had a strong 'DRG', the down-the-road-graphic – the visual impression the occasion of its passing creates. The first production 92 rolled off the line on 12th December, 1949. It bore a close resemblance to the 'design special' that was Ursaab. This is rare in car design, as most concept cars get diluted between prototype and customer product. But when a car is non-derivative of others, with pure design integrity, that tends to be very obvious. In the prototype variously known as Ursaab, Exprwagen or 92.001, Sweden's leading aircraft manufacturer produced a blend of elliptical ideas that reflected its recent past while suggesting an exciting future.

How did this happen? How did this aero-weapon of a car with Cd, in production form, of only 0.35 suddenly emerge? Under that airflow-tuned skin there was front-wheel-drive, a transverse engine, a flat floor, reinforced A-pillars and a built-in roll hoop. All this was years before the Mini and the Citroën DS earned plaudits for their respective claims to innovation.

Like other advanced cars of its era, the Ursaab was shaped by a sculptor, yet one who was also a pilot and an aerodynamics expert. His name was Sixten Sason. The 92 was called by the Swedes a *Vingprofil bil* – a wing-shaped car. Stylistically, Ursaab contained themes of Art Deco and Modernism with a touch of L'Esprit Nouveau. Surely Charles-Edouard Jeanneret, Le Corbusier, would have seen it as a piece of architecture that just happened to have wheels?

The Saab car idea stemmed from Saab's need to diversify as it became clear that the end of WW2 would bring a downturn in its business. Aircraft engineer Gunnar Ljungström was the true father of the Ursaab and its 92 offspring, and Sixten Sason the designer who shaped them at Ljungström's request. Rolf Mellde, a

talented engine and handling specialist, came to Saab just as the Ursaab, drafted in 1946, was emerging into its productionised form.

Ljungström was a time-served car fanatic and engineer who had worked for the British company A W Wickam in the 1930s. He had already designed wings and hulls for the Swedish aircraft maker. He was aware of DKW and, like Sixten Sason, had observed the work of Prof Dr-Ing Wunibald Kamm and Baron Reinhard von Koenig-Fachsenfeld, the two pioneers of 1930s road vehicle aerodynamics at the Stuttgart Institute for Vehicle Research. The Baron had already styled an aerodynamic fairing for a DKW racing motorcycle, one of the German products Saab was studying. Porsche's VW Beetle may have been teardrop shaped, yet it was an aerodynamic nightmare of airflow detachment, buffeting and ill-defined air-separation-point issues. Saab's teardrop, by comparison, was aerodynamic near-perfection.

Saab was unhindered, unblinkered by a previous psychology; Saab the car maker could start with a blank sheet of paper and new ideas. A harsh climate and extreme conditions in Sweden may have been relevant to the company's first car design, but they cannot have been the sole arbiter, otherwise Volvo would surely have been building something far more appropriate for the very same conditions.

DKW and a legacy
Saab was an aeroplane maker and research was vital. DKW's small cars had sold well in Sweden in the 1930s, so the starting point for Saab was to be a memory of a prewar German car. Key to the new project was Swedish car dealer and former DKW agent Gunnar V Philipson, who signed a contract for 8000 of the new cars and invested 1.8 million Swedish Kronor (SKr) in the project in advance.

Surely Saab, the aircraft maker, could take the basic idea of a car concept, apply its aircraft design and engineering ethos and create something more modern? That was the challenge.

The first design remit stemmed from a management decision that the Saab car should be both faster and more economical than the DKW. This could only be achieved through radically improved aerodynamic design and a much more efficient engine. Less drag meant more speed and less fuel. It was a simple equation, and thus was set the foundation of the Saab car ethos which was to last until the end of the fwd Moss-Carlsson era. Ljungström chose a teardrop, but one that was front-engined and front-driven with a valveless two-stroke that aped DKW's and Heinkel's. The engine was water-cooled and the car weighed just 765kg. The taught hull, low-roll suspension and buzzy 25bhp two-stroke delivered an unheard-of driving experience for the Swedes, with the result that the production 92 was an overnight rallying success.

The metal gauge used was more than 20 per cent thicker than on any contemporary car. The roof skinning was 1.11mm thick, the body side walls 0.87mm and the crucial, box-section steel sills were 1.59mm in thickness. In a unique, chassis-less aircraft fuselage inspired design, all apertures were reinforced, closely welded and riveted and the roof was a one-piece pressing with no leaded joints. There was no boot opening. Through such techniques exceptionally high torsional rigidity was achieved, reputedly at 11,300lb/sq ft in, an incredible achievement against a then industry average of 2000, and more recently 5000.

Thanks to Ljungström's previous role as a wing stress engineer at Saab, and the work of Olle Lindgren as stressman and of Erik Ekkers, the little Saab had a tough aircraft-style hull that resisted intrusion. A man named Bror Bjuströrmer, who was a figure in the Saab engineering group, has been cited as contributing to the initial schedule of design themes. Ljungström, Sason and project manager Svante Holm all guided the development of a wooden buck prototype at Saab's Linköping base prior to production at the new Trollhättan factory. Wood, horse dung, boot polish and the skills of a 70-year-old artisan woodworker were all employed to sculpt the elliptical form on a styling buck that Sason had created. For the new car, Erik Rydberg was the chief production engineer and tooling manager and Claes Sparre was factory engineer. Ursaab's test drivers were named Garbing, Nyberg and Landby. Gunnar Ljungström is reputed to have commented to those who were shocked by the car's suggested shape: "It doesn't matter if it has the look of a frog if it saves 100 litres of fuel a year."

The first car in the metal – built by a team of 15 men, some of who were just young apprentices such as Tage Flodén and Hans 'Osquar' Gustavsson – used second-hand parts, including DKW and Hanomag production items. An Auto Union fuel tank was fitted into the Ursaab prototype. Steel was in short supply and Saab had no previous parts bin to source from. But Ursaab was completed on 1st May, 1947, and a revised second version less than one month later. In 1948 Ursaab even sported a badge calling it the Sonett, a name Saab would apply to its composite construction two-seater sports car in 1955 and later to a 1960s range of models.

Aero-Auto-Mechanical
Ursaab had a Cd of 0.32 at a time when the industry average was closer to 0.50. Wind tunnel tests at the Swedish Institute of Technology showed that, despite lacking a Kamm-defined critical separation point, or bodywork ridge, Ursaab had good airflow characteristics. It even kept its airstream attached down its rump to reduce wake drag. Ursaab had a low lift coefficient and Sason's work on the boundary layer with smooth panels and small panel gaps allied to the windscreen angle and roof shape ensured no bubbling or detachment of air down the sides.

Studies in the 1970s by A J Scibor-Rylski were eventually to show that the 1940s 92.001's form, its windscreen rake angle and frontal curves or lobe approached the near-ideal aerodynamic form for a car.

Ursaab had to be lightly modified for production. The fully shrouded front wheels had to be partially exposed to permit U-turns and Alpine hairpins, and the wing aerofoil leading edge shape of the nose was too avantgarde even for Saab. Various proposals brought about twin podded lamps and a more upright grille, much to Sason's chagrin.

Was the front of Ursaab too bulbous, too excessive in cross-sectional drag? Scibor-Rylski's work on the ideal aerodynamic form indicates that the prototype was more intuitively correct than subsequent claims made about it have suggested. But for Ljungström's new car, Sason had walked into the office and put down a sketch of a raked, stylish, fastbacked teardrop that had shades of French and prewar Bohemian styling trends. Above all it reeked of the ellipse. Were elliptical specialists Lanchester, Voisin, Paulin, Shenstone, Joukowski, and

1. Genesis of Ursaab. This rare drawing of a 1936 Sason four-door aerodyne concept probably predates Ursaab by a decade

2. In this 1941 colour rendering, Sason is clearly hinting at the elements of his Ursaab proposal. Note the centre-line steering

3. This is Gunnar Sjogren's sketch of Ursaab as it was originally conceived with the faired-in front treatment

4. This side view and inset reveal the original Ursaab with its faired-in square headlamps and advanced scaling and sculpting

5. This is Sason's sketch number one for the Ursaab project, capturing what Ljungström wanted

6. An interim sketch for a more conventional front for Ursaab

7. The art of airflow. Tuft tests show that the domed roof, elliptical forms and smooth skin kept the airflow attached

Elliptical Excellence

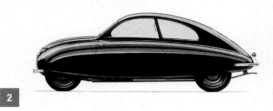

1. Unveiling Ursaab to the Saab management at Limkoping

2. Ursaab, known inside Saab as 92.001 or 'original Saab', seen in final prototype profile. It was not called a wing profile car for nothing

3. This is the production version of Ursaab, the 92. It is seen here with the Saab-designed hydrofoil boat at speed in the background

maybe even the obscure Gerin, deep within the substrata of the Ursaab's design consciousness?

The ellipse offered significant reductions in induced drag and cross sectional transitions, minimising the interference effect by enhancing curvilinear flow distribution where, normally, squared-off angles created turbulence. Parabolic, smoothed panelling delayed airflow separation. Saab the aircraft maker and Sason the pilot-designer would have known all this, but few other car makers did.

The roots of Sason's influences and styling preferences stemmed directly from his time spent studying sculpture and design in Paris circa 1936, and from his tours in Germany and Italy at the height of the Art Deco years. His knowledge came straight from early aviation and from the pioneers of aerodynamics. It was clear that the trends of the late 1920s and the Art Deco motifs of the 1930s, allied to tangents of thought from the likes of Paul Jaray, Jean Andreau, Gabriel Voisin and Hans Ledwinka, were also a strong influence. And what of Porsche and Joseph Ganz? Surely the spread of advanced aero-influenced experimentation taking place in Paris and Stuttgart at that time cannot have been anything other than inspiring.

More than any other contemporary design in 1945, Sason's work, his shape, his solo design style, was new. It was a Saab hallmark from the start – not a copy, not just a re-interpretation, but a defining shape that created a smooth yet compact coupé of a family car while the likes of Volvo, Morris, Renault and even Citroën were still churning out dinosaur designs.

But where did Sason's 92 shape come from?

Paris again?
The design influences upon Sason were essentially French, German, and latterly Italian and American. In the early 1930s he had gone to Paris to study sculpture. There he was thrown into the world of French streamlining. The principally relevant names of this era were those of Gabriel Voisin, Giuseppe Figoni and Georges Paulin whose talents touched not just French cars, but also the likes of the 1938 Bentley Embiricos (see Brian Sewell's article, page 162). In Sason's design of the headlamps and grille for the Ursaab derivative, the 1955 Saab 93, we can see very clear hints of Voisin's 1936 C27 Aerosport; the similarity of that powerful frontal graphic is obvious.

Sason also studied the art of working in silver, and a glance at the door handles of Ursaab reveals a design that reminds us of cutlery. The badges, too, seem extraordinary for what was a cheap, small car.

Saab, Citroën and sculpture
Also fascinated by sculpture in Paris at the same time as Sason was the Italian Francophile, Flaminio Bertoni (not Bertone). He was the man who went on to shape the Citroën DS after having previously sculpted the *Traction Avant* out of clay in a few days. In terms of modernity, perhaps the DS was the ultimate symbol of the design trend first set off by the 92.001. Certainly the 92 and DS seem to share a similar stance. Park one beside the other and the stylistic linkage is clear in terms of sculpture, scaling, angle, stance and domed form.

As we know (see *The Automobile*, October, 2011), Bertoni had to change the DS's rear and roof design at the last minute in late 1954 from a coupé fastback to a notched shape with a 'false' roof peak and ridge that sat six inches above his original sloping, metal profile. The reason for this was a claim given credence by Citroën that a German car maker was about to launch a similarly shaped car. No such challenger is known, but the wrap-around rear screen of the developed Ursaab/ 92, the Saab 96, was very close to the design Bertoni had intended for the DS. The shape was first drawn by Sason. Remove the DS's plastic roof and coroneted C-pillars and the riddle is perhaps resolved.

Ursaab's interior had shades of Bugatti design in its chrome framed seats, and of Citroën in the cabin fittings.

Ursaab underestimated?
If the observer sees the Ursaab as a private 'one off', its effect can be underestimated. Surely it was a stepping stone between Art Deco and the reverse engineering of post-war design austerity that followed it, to the beyond of a new design order. Ursaab opened the gates to cleaner, smoother, more integrated styling in the mass market – a quiet yet vital launching point for the more famous shapes that followed. And what of various ellipsoid American offerings that led to the use of aviation motifs in 1950s and '60s car design? Do these

Elliptical Excellence

4. The interior seems to have shades of Citroën, Bugatti and Art Deco themeology. Note the thick sidewall at the A-pillar

5. A DKW engine and other secondhand parts went under the bonnet of Ursaab. The fuel tank is reputed to be an Auto Union component

6. The final Ursaab prototype with round headlamps. Note the elliptical sculpting and ellipsoid structural elements

7. Sason's modified frontal treatment for Ursaab resulted in a more conventional visage for the flying saucer car

not all launch from a gear-change in design awareness that Ursaab and its Saab 92 offspring helped to create? Ursaab's progeny gained a massive profile in the USA, and three European royal families drove Saab 92/93 models as their private cars.

The last Saabs of 2011, including Jason Castroita's Phoenix design study with its re-interpretation of Ursaab's elliptical elements, alongside the Simon Padian-crafted shape of the short-lived Saab 9-5 Mk 2, contain clear elements reaching all the way back to a sculptural theme created by one man – Sason, in the 1940s. Saab, Porsche, Panhard, and Citroën designs are unusual in that sense of their creators' influence?

Saab has rarely received credit for its quiet contribution. Perhaps not being fashionably Italian was the explanation as to why the Swedes and that pioneering Saab think-tank have been obscured for far too long. In Ursaab belongs much of the afterwards and before of car design generally, as well as the very essence of the themes that defined Saab the marque almost up to its sad demise.

Bibliography
- Cole, Lance, *Saab Cars, The Complete Story*, Crowood Press, 2012
- Elg, Per-Borje, *50 years of Saabs: All the Cars 1947-97*, Motorhistorika Sallskapet Sverge, Sverge. Stockholm, 1997
- Scibor-Rylski, A J, *Road Vehicle Aerodynamics*, Pentech Press London, 1975
- Sjogren, Gunnarr, A, *The Saab Way*, Gust Osterbergs Tryckeri AB. Nyokoping, 1984
- Saab Veterarerna Arkiv
- *Grattis Saan 60 Ar!*, Hans O Gustavsson, Corren, Ostergotlands storsta, 8th June, 2007
- *Huruledes Grunden lades till Saabs bilproduktion vid flugmaskinsfabriken*, Hans Osquar Gustavsson and Siggvard Lenngren
- Personal correspondence: P Johansson

Sixten Sason –
thinking in three dimensions, styling in sculpture
Who was this designer and how did he create a new fashion with a small/medium sector car that looked like an expensive coupé?

Sixten Andersson (not yet Sason) was born in Skovde, Sweden, in 1912. It was not until the 1930s that, as an emerging industrial designer and stylist, he altered his last name to Sason – alluding to spice and less common than the ubiquitous Andersson. The first name Sixten derives from the old Swedish words *sighan*, meaning victory, and *sten* (stone). But Sason was no flamboyant artist. He was a serious-minded engineer, a calm thinker who could work with men like Ljungström and Mellde. Ljungström said of him: "A genius; an engineer with the talents of an artist, or an artist with the temperament of an engineer...the ideal partner to work with."

Tall, handsome, a debonair and charming Swede with a taste for Italian style, Sason was a pilot and a technical thinker. His 1930s ideas for a micro-car predated the post-war German designs from BMW, Heinkel and others by more than a decade. Other early designs included a people carrier and an aerodynamic, low slung steam-powered limousine.

The observer might argue that the *goutte d'eau* or teardrop school of design was not new, but Sason's interpretation of it was novel yet timeless – certainly not a quickly dating fashion. The Saab 92 was an amalgamation of ideas and themes that appealed not as a retro-pastiche, but as something genuinely new.

And the 92 was not Sason's only defining work. From an early fascination with flying, the then Sixten Andersson trained as a pilot in the Swedish Air Force. In a subsequent crash his chest was penetrated by a wing strut, he lost a lung and was struck down by infection. He was in hospital for months, yet within a few short years he had re-invented himself as an illustrator and then as an industrial designer. He also studied silversmithing, and that may explain some of the exquisite detailing in his other designs, be they cleaners, cookers or cameras.

By 1939 Sason had started to work in the illustration department of the rapidly expanding aircraft maker that was Svenska Aeroplan Aktiebolaget – or SAAB, as Swedes termed it. Initially, he specialised in producing X-ray-type see-through structural drawings of Saab aircraft and their components.

He also designed the Husqvarna Silver Arrow motorcycle in 1955, the Hasselblad camera of 1948 and the company's later 1600F series. Sason shaped the first 1940s Electrolux Z70 vacuum cleaner, the Monark moped, and not only a Husqvarna chainsaw but also that manufacturer's waffle cooker – a stunning device that looked just like a spaceship or UFO beamed in to Area 51. Or indeed to Hangar 18 at Wright Patterson's secret air force base, the place where in the war years Alex Tremulis had sketched and styled the flying saucer-shaped devices for the US military long before he designed cars that included the Tucker...

Sason also styled a range of cookers and fridges, and came up with a defining shape of electric hand-iron – one copied even today. He designed the Zig-Zag sewing machine, and his drawing board and studio were packed with sketches for boats as well as elliptically-shaped flying machines of great futurism. Sason even worked on a proposal for a curved-buttress suspension bridge at Oresund.

All this work was textbook product design and industrial design, function and perfection without excess. Sason's designs were integrated, mature, organic and original. In a sketch of 1941, he shaped a delta-winged, all-wing or blended-wing short take off and landing rocket-powered fighter that, more than 70 years later, still looks like an advanced stealth design. How much this 1941 design influenced Saab's later Erik Bratt-designed Draken jet fighter, itself a design icon, is a long-argued point, but there are those, including Rolf Mellde, who have suggested some similarity, so maybe a link is not as fantastical as some might argue.

Sason's archives, exhibited in the Vastergotland Museum in 2009, also show that in the 1930s he was designing micro-cars that pre-dated the post-war German fashion by more than a decade. He drew a people-carrier MPV and shaped the Catherina two-door styling proposal that introduced the targa-top roof design and had flying C-pillar buttresses – both ideas later used by other designers. He played a role in shaping Rolf Mellde's Saab 94 Sonett project.

Sason had made a study of Werner von Braun's advanced rocket missiles. He had even sketched the details of a crashed Nazi V-1 rocket and then, it is reputed, travelled to Britain in the wartime bomb bay of the clandestine BOAC de Havilland Mosquito service to present his findings to British Intelligence and their Air Ministry boffins.

Sason was known for his long research trips round the design houses of northern Italy, where he was held in great affection. He died in April, 1967, aged only 55, just as his last car, the Saab 99, was about to be launched. By then he had taken on the young Björn Envall and encouraged him to contribute to the 99 as his own health failed. Envall, himself the father of 1970s-80s Saab design, relayed to the author that "Sason had a strong sense of humour; he was quite English in that sense. He was the best, a top industrial designer."

Sixten Sason died tragically early, but left a legacy of elegance and function. His Husqvarna motorcycle and his waffle cooker symbolise his blend of styling and functional elements at their best. But perhaps his flying saucer of a car, the Ursaab 92.001, and its production descendants leading up to the 99, should qualify him as one of the unsung heroes of 20th century automobile design.

Whatever the verdict, Sixten Sason was a true renaissance man, a forgotten hero of design whose cars touched the soul, and whose Hasselblad camera design went to the moon. How many designers can say that?

1-2. The Saab Super Sonett I, or Saab 94. From 748cc it could reach speeds of 120mph

3. Sason's late 1930s limousine proposal, with glassed dome roof and faired-in wheel pods

4. This Sason x-ray sketch of the Saab 92 structure formed the basis of the Västergötlands museum's publicity for its Sason tribute in 2008

5. Sixten Sason, or Karl-Erik Sixten Andersson as was. One of the 20th century's outstanding industrial designers, yet little known outside his homeland

6. In the early 1960s Sason drew this idea for a two-door coupé, in the author's view much better looking than the later Saab Sonett II and III

7. Another rarely seen Sason image, this time of a smaller Saab for the 1960s-70s apeing the 99's styling elements

J.D Classics

Office: (01621) 879579 *Facsimile:* (01621) 850370 *Mobile:* (07850) 966005

AUSTIN HEALEY 100 S
**WORKS RACE PREPARED IN 1954 FOR AUSTIN HEALEY CANADA.
PERIOD COMPETITION HISTORY, FOUR OWNERS SINCE NEW.
FULL CHASSIS OFF RESTORED, RACE PREPARED BY JD CLASSICS.
RACED AT LE MANS CLASSIC 2012, SUBSTANTIAL ORIGINAL SPARES PACKAGE.**

WEB SITE: www.jdclassics.co.uk
EMAIL: jdclassics@jdclassics.co.uk
OFFICE: (01621) 879579 FACSIMILE: (01621) 850370
MOBILE NUMBER: (07850) 966005 or (07860) 824531

WYCKE HILL BUSINESS PARK, WYCKE HILL, MALDON, ESSEX CM9 6UZ, U.K.

A Grand Reunion

Keith Bluemel describes a Mille Miglia-winning Ferrari and was there when it was reunited with its original owner, Count Giannino Marzotto

The 166 MM Barchetta with coachwork by Carrozzeria Touring of Milan is the model that established Ferrari on the world stage, at the same time giving the marque an identity. Its elegant yet purposeful lines have captivated enthusiasts since it first appeared at the Turin Salon in 1948. The model went on to project the Ferrari name to an international audience through wins at the Le Mans 24 Hour and the Spa 24 Hour races in 1949, and the egg-crate grille first featured on this model provided a visual identity that continues to this day. In both of these races part of the driving team was Luigi Chinetti who, apart from his driving skills, had sales skills. He was an important link in selling Ferraris commercially, particularly in the US market which was recovering more quickly from the effects of WW2 than its European equivalents.

The 166 MM Barchetta by Touring had a closed roof sibling, the 166 MM Berlinetta Le Mans, of which only six examples were produced by the Milanese carrozzeria. They featured a near identical body style below the waist line, but incorporating a full-height split windscreen with wipers, and a roofline that flowed in a majestic sweeping curve into the lower tail section. Sliding windows were provided in the doors, and the rear wing line was raised so there was a valley between its surface and the roof. This swept in a graceful arc beneath the circular tail lamps in the wing extremities, a feature also found on a number of late production 212 Touring Barchettas as well as on Alfa Romeos of the period and on the Touring-bodied Bristols. The Le Mans appellation in the Ferrari title was a reference to the 1949 Le Mans win, in much the same way that the MM (Mille Miglia) part of the 166 MM model name was a reference to the 1948 Mille Miglia victory by a 166 Sport.

The six berlinettas were produced on chassis numbers 020 I, 0026 M, 0042 M, 0048 M, 0060 M and 0066 M. The car featured here is chassis number 0026 M – by chassis number the first true 166 MM Touring Berlinetta, and probably the first one produced, as chassis 020 I, although an earlier number, was originally a 166 Spider Corsa chassis. Research undertaken by David Seielstad suggests that chassis number 020 I was part of the 166 Spider Corsa programme, modified by grafting on the rear section of a 1950-type 166 chassis from the transmission location rearwards to make it, possibly, the prototype for the 1950 chassis series. It was sold as a rolling chassis to Franco Cornacchia in 1949, who sent it to Carrozzeria Touring to have a berlinetta body fitted. It is a reasonable supposition to think that he saw what they were producing on 0026 M, and decided to follow the same route.

Chassis number 0026 M was sold new to Giannino Marzotto, the third of five sons of Count Gaetano Marzotto, four of whom, Vittorio, Umberto, Giannino and Paolo, were proficient gentleman racing drivers (only the youngest, Pietro, didn't follow his brothers down that road). They owned a number of Ferraris in the early '50s, having raced predominantly in Lancia Aprilias in the late '40s. However, their father disapproved of them racing those 'red cars', so they frequently had them painted in other colours, and were not averse to changing bodies on occasion. Thus 0026 M was finished in blue. Some sources say this was in recognition of the win at Le Mans the previous year. The Certificate of Origin for 0026 M was issued by the factory on 2nd February, 1950, but according to the book *La Saga dei Marzotto* by Cesare de Agostini, published by Nada Editore in 2003, when Giannino tested the car for the first time around Maranello he was disappointed with the performance and expressed his feelings to Enzo Ferrari. The car must have remained at the factory after this first encounter, as it was displayed at the Geneva Salon that March.

It is understood that, unbeknown to Giannino Marzotto, the engine was changed for a larger capacity 195 unit (2.34 litres) prior to his drive in the Mille Miglia in the May of that year. According to the same source, the lack of power had mysteriously disappeared by that time. He was advised that he was a 'works' driver, and that the car now sported a 2.3-litre engine.

He certainly used the additional power to good effect. With co-driver Marco Crosara he won the race, with the 195 S Touring Barchetta, chassis 0038 M, of Dorino Serafini/Ettore Salami finishing second. The winning car was subsequently displayed, still with the grime it had accumulated during the Mille Miglia, at the Turin Salon in May, 1950.

Thus the 195 Sport models were born. They were essentially larger-engined versions of the equivalent 166 MM Touring Barchettas and Berlinettas. In fact, factory records show that the examples 'produced' were all originally earlier 166 MM cars upgraded to 2.34-litre engine capacity, this work being carried out in 1950-51. The 166 MMs noted as having been converted to 195 specifications were Barchetta chassis numbers 0016 M, 0038 M, 0040 M, 0044 M and 0052 M, whilst the Berlinettas were 0026 M, 0060 M and 0066 M, although it is possible that other examples may have been recipients of the upgrade. The V-12 engine was increased in capacity from 1995cc to 2431cc by increasing the bore from 60mm to 65mm whilst retaining the 58.8mm stroke of the 166 engine.

As an aside, the 1950 factory sales brochure, *30 Anni di Esperienze*, shows an artist's rendering taken from a photograph of the Yvonne Simon/Michel Kasse 166 MM Berlinetta, chassis number 0042 M, raced by them at Le Mans that year (possibly the first post-war female pairing in the race), notated as a 195 Inter model.

On the subject of Le Mans, 0026 M was another participant in 1950, appearing with a long intake scoop in the bonnet and driven by Sommer/Serafini. It led for the first two hours, setting a new lap record, before spark plug problems and then a failed dynamo bracket saw it posted as a retirement 10 hours into the race. Between the display at the Turin Salon and Le Mans, it had competed at the Coppa della Toscana where, driven by Franco Cornacchia, it finished second. Back in the hands of Giannino Marzotto it then won the Tre Ore Noturna (Three Hour Night Race) at the Caracalla circuit in Rome. He removed the engine from 0026 M and fitted it in the family's 166MM Touring Barchetta, chassis number 0034 M, to win the fourth Coppa Internationale delle Dolomiti in July, 1950.

The car was eventually officially road registered in December, 1950, and then sold to a cousin, Baron Domenico Rossi, in January of 1951, going through a further two Italian owners before being exported to France in the mid-1950s. From there it went to the USA with various owners over the years before being sold to Switzerland in 2001, where it remained for only a short time, once again crossing the Atlantic to its current owner, Jack Croul, in 2003. He commissioned Paul Russell to undertake a complete restoration of the car back to its 1950 Mille Miglia-winning configuration, which took three years. The result, as can be seen from the photographs, is truly spectacular. Since its completion the car has appeared at numerous shows, and has been back to its old hunting ground, the Mille Miglia, to participate in retrospective runnings in the hands of its owner. In 2007 it won the Mille Miglia Trophy at the Pebble Beach Concours d'Elegance, and in 2008 it won Best of Show awards at both the Cavallino Classic and at the Villa d'Este Concorso d'Eleganza.

1. The Mille Miglia-winning 166M holds its own against the stylish formal backdrop of Count Marzotto's garden

2-5. Compact, light, and with a lusty V-12 under the bonnet, the little Ferrari had few rivals on Europe's early post-war racing circuits. That it was also very beautiful only adds to its appeal to today's collectors, who cheerfully part with seven-figure sums on the rare occasions when a well provenanced survivor comes on the market

General technical specification

Engine	60deg V-12, longitudinal, alloy block and cylinder heads	Front suspension	independent with transverse leaf spring and Houdaille hydraulic shock absorbers
Timing gear	two valves per cylinder, single overhead camshaft per bank	Rear suspension	rigid axle, semi-elliptic leaf springs and Houdaille hydraulic shock absorbers
Bore and stroke	65 by 58.8mm	Steering	right-hand drive with worm and roller steering box
Unitary/total displacement	195.08cc/2341cc		
Compression ratio	8.5 to one	Wheels and tyres	front: 72 spoke wire wheels, 15 by 4in with Rudge hubs and 15 by 5.50 tyres; rear: 72 spoke wire wheels, 15 by 4in with Rudge hubs and 15 by 6.00 tyres
Maximum power	160-180bhp at 7000rpm		
Induction system	one or three Weber 36 DCF/1 carburetter(s) with mechanical Fispa fuel pump		
Ignition system	twin coils and distributors	Brakes	400mm diameter drum brakes front and rear, hydraulically-operated, with cable handbrake to rear wheels
Electrical system	12-volt, with dynamo and Marelli battery		
Transmission	dry single-plate clutch, five-speed-and-reverse gearbox	Wheelbase/front/rear track	2250mm/1278mm/1250mm
		Kerb weight	850kg (approx)
Chassis	elliptical-section steel tubes	Fuel tank	85 litres

The Grand Reunion

An Audience with the Count

It is not every day that one gets the opportunity to mix with the aristocracy when one lives in an unfashionable suburb of London, much less someone with a motor sport heritage second to none. Thus it was a surprise to get a telephone call from a friend in Italy, Gabriele Artom, enquiring whether I would like to accompany him, and the 1950 Mille Miglia-winning 195 Sport Touring Berlinetta, to visit the original owner and winning driver, Count Giannino Marzotto, at his home to reunite him with a car he hadn't seen for nearly 60 years.

Of course the answer was in the affirmative, the Marzotto brothers being legendary gentleman Ferrari drivers in the company's early years. The car had been away from its current California home, in Italy for the retrospective Mille Miglia that May, and Gabriele was 'babysitting' for the owner, Jack Croul, prior to getting it shipped home. He thought it would be nice for the Count to meet up with his Mille Miglia-winning car again. Gabriele's father, Franco Artom, initiated the introduction as a long-time friend of Marzotto, both from his youth and in later years through business and sporting pursuits. All I had to do was arrange a flight to Milan to meet up with Gabriele, and he would organise everything from there – job done!

The flight and the train journey into Milan constituted the relaxing part of the trip. As anybody who has driven with Gabriele will confirm, whether it be in an old car or a modern one, he only has one speed – flat out, a real Italian trait. Fortunately for the 195 it was being transported by truck. We were in a modern BMW, where at least I had seat belts and airbags should the worst come to the worst. Almost unbelievably we made it to the Villa Marzotto in Trissino without mishap, and as far as I am aware without a speeding ticket.

We arrived just as 0026 M was being disgorged from the truck in the street at the entrance to the Villa's grounds, so Gabriele's manic driving proved to be perfect timekeeping. A quick mobile phone call to the Count's PA to advise of arrival and we took the car up the driveway to the imposing villa, where she greeted us and said we were free to use any locations we wished to take pictures until the Count came up from the 'small house' on the estate down the hill on the other side of the town. The date had been arranged to coincide with a party that he hosts annually for locals in colonnades around a courtyard adjacent to the 'small house', so we were invited to sample his hospitality for the evening. She mentioned that he had initially appeared ambivalent to the idea of the reunion, but seemed more hospitable to it as the date approached. She confided that he normally took a nap after lunch, but hadn't done so today, so it seemed the prospect was giving him a degree of excitement.

After taking a multitude of photographs, we had to decide where to place the car for his arrival. Here the weather came into play, as it started to rain steadily, which wasn't the best scenario for the presentation. Another call to the PA. She suggested an outbuilding attached to the villa where there was space to house the car in the dry, which would give us the opportunity to dry it off before he arrived. A further suggestion was that maybe we would like to have the winner's cup from the Mille Miglia with the car for the presentation, and she duly arranged that. She showed us the way to a pair of large grey metal doors which appeared not to have been opened for years, as they creaked on their hinges upon

1. Reunited after 60 years, Giannino Marzotto and the car that powered him to a memorable win in the Mille Miglia of 1950 pose with the victor's trophy, still in the Count's possession when he died, aged 84, earlier this year

opening. There was a large clear area just inside the cavernous unlit space, the only light coming from the door opening, where in a corner in the gloom lay forlornly a 1960s Iso Rivolta on axle stands, covered in a thick layer of dust, obviously forgotten and untouched for years. As we finished drying off the Ferrari, she received a phone call to say that he was on his way up from the 'small house', so we hurriedly closed the doors to hide the car from view and stood in the, now fortunately light, rain awaiting his arrival.

A silver Mercedes-Benz estate car swept up the driveway to the building and out stepped the Count, cigarette in hand, a tall, powerfully built, imposing figure, not at all like the jockey size race drivers of today. One immediately had the feeling this was a man of substance and power, even though he was in his 80s and had been in poor health shortly prior to our visit. He had an indefinable presence, and exuded an air of grace and dignity as he greeted us with a firm handshake and a warm, welcoming smile.

After the general introductions, Gabriele cheekily commented on the large ashtray sitting atop the regular ashtray in his car, brimmed with cigarette butts. He asked him whether he smoked when driving in the Mille Miglia, to which he replied in the affirmative, adding that it was dangerous but he still did it. "And now I smoke as much as I can," he added with a wink. The greeting smile broadened into a wide grin when we swung open the doors to reveal his 'mistress' of nearly 60 years before, still radiating her seemingly eternal youthful beauty. Any misgivings he may have had about a reunion must have dissipated immediately upon setting eyes on her gorgeous form, as he talked animatedly (in Italian, so much of it was lost on me until Gabriele translated later) whilst walking around her and caressing her curvaceous body. He slipped inside to grip that slim wood rimmed steering wheel once again, and casting his eyes across the dash panel he immediately noted a gauge suspended below it, saying that it wasn't there when he owned the car and that the interior

The Grand Reunion

2. Made to measure. Despite his 80+ years, Count Marzotto slips behind the wheel of his old friend and prepares to drive it across the park to an evening gathering of friends

mirror was lower, also commenting that there was no way you could heel and toe with the pedal set-up as it was. What a memory for an octogenarian who hadn't seen the car for close to 60 years!

By now the rain had ceased, so we were able to bring the car out of the gloom into the hazy sunny afternoon, whereupon he noticed something else that he didn't recall being on the car when he owned it: the vertical extractor slots in the perspex rear side windows. Of course, once again he was right; they were added by Jack Croul to aid cabin ventilation, although similar slots had been a feature of some cars in the series. After a visit to his office in the villa, where he showed us some of the trophies he had won during his racing career – things he had kept without any great thought, as he said he always looked forward and didn't really reminisce. Only in recent times had he given any thought to the past, but mainly as a legacy for the family, not for his own gratification.

Gabriele reminded him about the surprise of the people when he took the chequered flag upon winning the 1950 Mille Miglia, impeccably dressed in a double breasted suit and tie. At this he laughed loudly, saying he normally raced in overalls but always kept a suit in the car in case he broke down and had to take the train home. His family wouldn't think it befitting for a Marzotto to travel in public in grubby racing overalls. A further question Gabriele posed was about the single versus triple carburetter set-up, to which the Count replied immediately, without hesitation, that it was initially tested with a three-carburetter arrangement, but reverted to a single Weber – another testimony to his undimmed memory. We realised that he had been seriously interested in the cars he drove, and not just a proficient wealthy amateur. Prior to leaving his office he kindly wrote a short dedication in my copy of the book *La saga dei Marzotto*, which I had brought with me. Loosely translated, it read, 'We don't have to race to win, but to compete, as competition produces progress, which is something that I believe in.'

This reinforced the philosophy he had expressed earlier, of looking forward and progressing – definitely a man in control of his life, and with a sense of direction and purpose.

He invited us down to the 'small house' for an aperitif prior to the evening's gathering. As we sat around a table in the garden, we noticed a sculpture in the distance. It was actually a badly crashed Iso Rivolta, in which he had had an accident in the '60s, wrapping it around a tree with such force that the driver's footwell was reduced to almost nothing. He had walked away virtually unscathed, and the wreck stands as a reminder to him to be more careful henceforth. He also recalled an incident from his youth, pointing to the large field that adjoins the house and garden. Evidently it was used as a landing strip for light aircraft, and he had started to learn to fly. His flying career came to an abrupt end when he crashed through the kitchen wall after landing, and his parents banned him from the skies.

During the chatter round the table, further proof of his technical capabilities, and skills as a driver, came to light. It emerged that he was loaned prototype test cars by Mercedes-Benz to evaluate and provide technical feedback prior to production, one of these being the A-Class, which he told them needed the suspension sorting out. We all recall the infamous rollover when tested by a Swedish magazine in 1998, which sent the model back to the drawing board.

Soon it was time for the party to start. He used the beautiful blue Berlinetta to make his entrance, driving from the 'small house' across the lawn through to the courtyard, where the gathered locals gave him a round of applause as he blipped the throttle, giving everybody a wonderful V-12 crescendo of sound to get the party started. He had a couple of final comments on 0026 M after his short drive: the brakes pulled to one side, and if the differential had been making the noises it was today, then he wouldn't have finished the Mille Miglia, let alone won it. But then that's the difference between the concours circuit and the other kind...

Vintage & Prestige
fine motorcars

Viewings By Appointment Only.
Unit 9b, Westfield Farm,
Medmenham, Nr Marlow,
Bucks, SL7 2HE

01442 236711
07967 260673
www.vandp.net

1903 Curved Dash Oldsmobile. 4 seat London to Brighton. £48,000

1912 Renault 5 litre Limousine by Kellner et ses fils. £138,000

1913 Rover 12hp four cylinder coupé. Time warp condition £39,500

1913 Delage Ai Boat tail Skiff. 50MPH crusing. £62,500

1912 4.0 litre Sunbeam tourer. Concours restoration. £168,000

1913 6.5 Litre Sunbeam tourer. Superb condition. £240,000

1924 Bentley 3.0 Boat tail. £40K Just spent. £185,000

1922 Silver Ghost totally restored London to Edinburgh. £240,000

1923 Daimler 9.4 Litre Hooper Limo. 1 of the Royal 7. £205,000

1925 Citreon Trefle. Gallic charm at its finest. £12,700

1926 Rolls Royce 20hp Barker Cabrio. Scumbled wood body. £79,000

1931 Humber 80 Snipe 3.5 Litre 6 cylinder tourer. £44,500

1929 Rover Weyman saloon. Total restoration. £17,900

1934 Alvis Speed 20 SB tourer by Martin & King. £108,000

1936 Alvis Speed 20 SD tourer. Huge expenditure. £100,000

1937 Lancia Aprillia Berlina pillarless saloon. £36,500

1937 Riley Big Four Redfern tourer by Maltby. £48,000

1939 Rolls Royce 25/30 Park Ward Swept Back. £42,500

1934 Stutz Vertical Eight. Last Built. English Body. Concours. £98,000

1979 Jensen Interceptor Mark III. Massive restoration. £29,000

1965 Mercedes 220SE convertible. LHD. Drives beautifully. £28,500

1950 Jaguar Mark V 3,5 LHD 3 position drophead. £74,000

1980 Bentley MPW Coupe. One of the last 7 made. 50K miles £52,500

CARS WANTED!!!
We will Buy your Car for Cash or Consign it.

FREE VALUATION SERVICE
Ask For: Richard Biddulph

THE ENGLISH MODEL T FORD
A Century of the Model T in Britain

A major work including much unpublished material and information collected from private and public archives, including Trafford Park, Manchester factory records, and Ford Dealers' sales records.

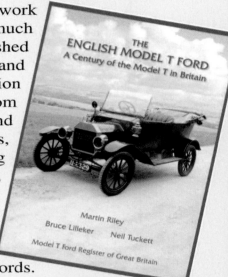

'A model of what a single model history should be. Beautifully produced on art paper and copiously illustrated with contemporary photographs.' — *Michael Worthington-Williams*

Hundreds of period and colour pictures.
304 Pages. A4 size. Hardback with dust jacket
Price £29.00 — including UK Postage and packing*
*For Overseas Airmail please add £5.00 for Europe. £11 for Rest of World
For details and to buy on line: **www.modeltbook.com**
Published by Model T Ford Register of Great Britain
Upper Holt Farm, Barber Booth, Edale, Hope Valley, S337ZL
Tel: 01433 670420   sales@modeltbook.com

Tim Walker Restorations Ltd
Buckinghamshire (15 mins J9 M40)
Established 1980

Servicing, repair and award winning restoration of Veteran, Vintage, PVT and Classic motors since 1980.

A small family business based 15 minutes drive from Aylesbury and Junction 9, on the M40.

Please call us for a chat on 01296 770596
www.timwalkerrestorations.co.uk
Email: guytw@aol.com

classic tyres for your classic
authentic tyres from the ultimate tyre authority

Established in 1962 we are now a truly global supplier, sourcing from 25 different countries and supplying to over 40. From motorcycle, car and race ...to commercial, military, aircraft and even industrial tyres, all reassuringly backed by our expertise. And, as you can imagine, such a worldwide company has the power to offer you excellent prices.

- FREE DELIVERY ON UK ORDERS OVER £50
- HUGE RANGE AVAILABLE FOR SAME DAY DESPATCH
- ALWAYS COMPETITIVELY PRICED
- FRIENDLY, EXPERT ADVICE FROM CLASSIC ENTHUSIASTS
- KNOWLEDGEABLE WITH YEARS OF EXPERIENCE
- COMPREHENSIVE RANGE OF ORIGINAL EQUIPMENT TYRES
- OFTEN COPIED, BUT OUR TYRES ARE THE REAL THING

- ALL OF OUR TYRES COMPLY WITH EUROPEAN LAW
- CONSTANTLY SEARCHING FOR ORIGINAL MOULDS TO DEVELOP 'NEW' ORIGINAL PRODUCTS
- BASED AT THE NATIONAL MOTOR MUSEUM, THE HEART OF CLASSIC MOTORING WITH UNRIVALLED RESOURCES

No comparison really.

- DUNLOP
- MICHELIN
- WAYMASTER
- Firestone
- VREDESTEIN
- ENSIGN
- AVON

CELEBRATING IN 2012 — 50 YEARS

vintagetyres.com
vintagetyres.fr
sales@vintagetyres.com

t: 01590 612261
f: 01590 612722

VINTAGE TYRES

THE BEAST OF TURIN

Stefan Marjoram shares some of his photographs and illustrations documenting the rebuild of the FIAT S76, the 28-litre record car which will reappear in 2013

1. Although the FIAT looks enormous from the side, its designers knew all about keeping the frontal area as small as possible

2. The engine is from the second car – FIAT scrapped the rest of it

FIAT S76 – The Beast of Turin

3. The car with its cowl fitted. The only way to really see if the body is right is to place it on the chassis, stand back and compare it to the photographs

4. The gearbox had to be made from scratch using original drawings

5. Some of the gearbox parts look quite delicate, considering the engine will put out 300hp at 1000rpm, with 2000lb ft of torque

6. The camshaft can slide to decompress the engine for hand-starting

1. As well as photographing the car's rebuild, I've been drawing it at various stages, too

2. The wooden former for the shapely tail...

3. ...and after it has been skinned

FIAT S76 – The Beast of Turin

4. Some original FIAT factory drawings have proved invaluable during the rebuild

5. Whilst test-fitting the seats I got to experience the car from the driver's position. Peering down the side of the bonnet gives one an alarmingly restricted view

Looking at old photographs of Land Speed Record cars, I've always felt it would have been wonderful to experience those scenes first-hand. Not just the moment when a successful team stood proudly behind their winning car, but also the earlier moments in the workshops when the car was being built. I can't remember rubbing any brass lamps but, amazingly, in 2010 I got that exact opportunity when I joined the *Bloodhound* SSC Land Speed Record team as its filmmaker, photographer and artist. At the same time I met Duncan Pittaway who, after years of searching the globe for remaining parts, had finally begun assembling the FIAT S76. I now have the enviable task of recording in photographs and drawings the birth of these two astonishing record cars, separated by exactly 100 years.

In 1910-11 FIAT built two S76 racing cars, perhaps partly with a view to getting one over on their long-term rival, the 200hp Benz. Also known as the 'Beasts of Turin' or 300hp FIATs, at just over 28 litres their four-cylinder engines were the largest car units ever built. At first glance the cars themselves look so utterly unusual to our eyes – almost comic – but they actually feature some very advanced ideas for the time. The S76s had a brief but action-packed few years, appearing at Brooklands, Saltburn and Ostend with drivers Arthur Duray, Felice Nazzaro and even a Russian prince before apparently disappearing at the time of WW1 and the Russian Revolution. One car did achieve an official flying-mile record of 116mph in less than perfect conditions, but it was clearly capable of much more. In the almost 100 years since they vanished, many tall stories and inaccurate articles about them have appeared. I'll leave it to Duncan to try to tell the full story in these pages later.

Sitting sketching the FIAT in the workshop has been a real treat. I've enjoyed being shown the beautiful craftsmanship and all the tiny details, such as the stamps and engineers' marks which will eventually all be sealed up inside the engine. It has been interesting to see on large photographs how the cars were altered during their short histories. Duncan blows each photograph up to enormous size and scrutinises every detail, for although he has wonderful full-scale assembly drawings, the finished cars were quite different. A studio photograph of the car at Brooklands has recently been found in a skip; it's very clear and has provided some excellent information on how the mighty FIAT was put together. I've found it fascinating to watch how a car of this period was constructed, from Bruce's beautiful timber frame that supports the body to the body itself. Last week I even got to help guide the cowl through a bead roller.

It's a steady process, as is the building of *Bloodhound*, but Duncan is determined to get the car right in every detail. Most of the effort at the moment is going into the bodywork. Once that's done and everything has been aligned, the whole car will be dismantled again before being painted and reassembled. I think I'll be about as excited to hear that huge engine fire again as I was at our recent *Bloodhound* rocket test.

For the most original & rare SS - Jaguar and Aston Martin cars

WWW.ZWAKMANMOTORS.COM